AF567168

Practical Interventional Radiology of the Peripheral Vascular System

Practical Interventional Radiology of the Peripheral Vascular System

PRACTICAL

Interventional Radiology of the Peripheral Vascular System

edited by

ANNA-MARIA BELLI

Edward Arnold

A member of the Hodder Headline Group

LONDON BOSTON MELBOURNE AUCKLAND

First published in Great Britain 1994

Distributed in the Americas by Little, Brown and Company,
34 Beacon Street, Boston, MA 02108

British Library Cataloguing in Publication Data
Belli, Anna-Maria
Practical Interventional Radiology of
the Peripheral Vascular System
I. Title
616.1

ISBN 0-340-55865-2

Photoset in Linotron 202 Bembo by
Rowland Phototypesetting Limited, Bury St Edmunds, Suffolk.
Printed and bound in Great Britain for Edward Arnold, a division of Hodder Headline PLC, Mill Road, Dunton Green, Sevenoaks, Kent TN13 2YA by Butler and Tanner Ltd, Frome, Somerset.

Preface

The number of interventional vascular procedures performed in radiology departments has steadily increased over the last decade, and these are no longer confined to specialist centres. Such procedures are often performed in preference to surgery, and surgery reserved for those in whom percutaneous methods fail, or are inappropriate.

The purpose of this text is to provide practical guidelines to those relatively inexperienced in the field and those who perform these interventional procedures intermittently. The opinions expressed in each of the chapters are those of the authors, all of whom have confined themselves to their own experiences to provide a sound basis on which the reader can build.

All the procedures described here are standard. Newer, experimental interventional vascular procedures such as lasers, atherectomy devices and stents, are not included as they are still undergoing evaluation and frequent technological changes and any texts including this information will be rapidly outdated. For this information the reader should refer to current journals.

The rapid expansion of interventional vascular radiology makes it one of the most exciting fields in which to be involved. It is hoped that by providing simple basic guidelines, the reader, having obtained the appropriate diagnostic vascular experience will feel confident and competent to proceed with the support of clinical colleagues to the simple interventional procedures. Experience in the basic techniques, knowledge of where to seek advice, attendance at interventional meetings and training visits to recognized specialist centres should enable the reader to proceed confidently to the more complex procedures and use new devices and methods where appropriate.

Anna-Maria Belli
1993

Contents

Preface v

Contributors ix

1 Percutaneous transluminal angioplasty 1
Anna-Maria Belli and David C. Cumberland

2 Thrombolysis 29
Roger H. S. Gregson

3 Embolization 61
Anne P. Hemingway

4 Retrieval of intravascular foreign bodies 81
Anna-Maria Belli and Anne P. Hemingway

5 Inferior vena cava filters 93
Irving Wells

Index 119

Contributors

Anna-Maria Belli, DMRD, FRCR,
Consultant Radiologist, St George's Hospital, London

David C. Cumberland, FRCP(Ed), FRCR,
Consultant Cardiovascular Radiologist, Northern General Hospital, Sheffield

Roger H. S. Gregson, DMRD, FRCR,
Consultant Radiologist, University Hospital, Nottingham

Anne P. Hemingway, FRCP, DMRD, FRCR,
Senior Research Fellow and Honorary Consultant Radiologist, Hammersmith Hospital, London

Irving Wells, MRCP, FRCR,
Consultant Radiologist, Derriford Hospital, Plymouth

CHAPTER 1

Percutaneous transluminal angioplasty

Anna-Maria Belli and David C. Cumberland

Indications 2

Contraindications 2

Patient preparation 2

Equipment 3

Arterial Access 8

Recanalization technique 16

Postprocedural care 21

Complications and their management 2

Results 24

New devices in angioplasty 25

References 26

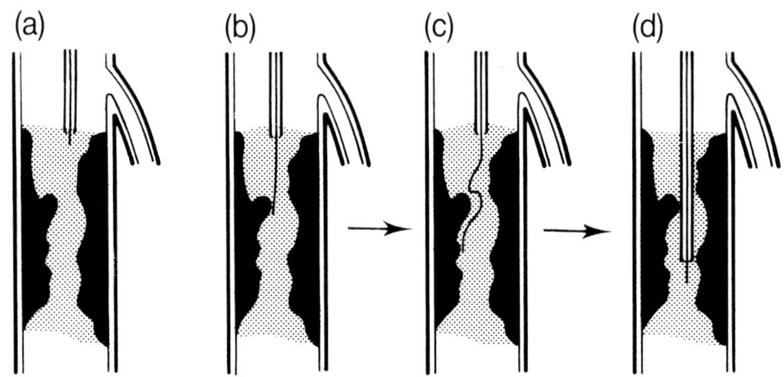

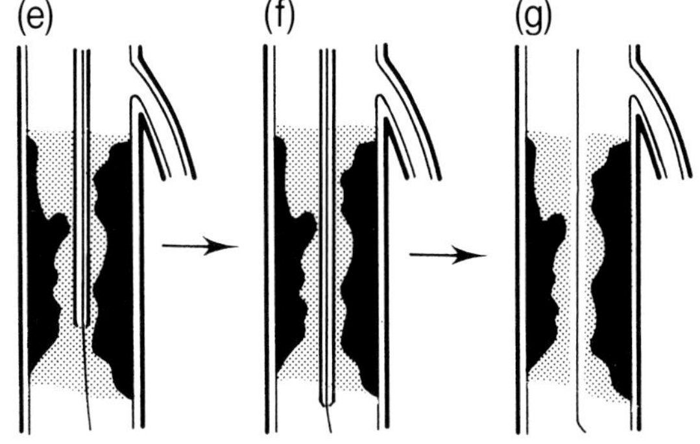

Indications

Patients with peripheral vascular disease are graded according to the severity of their symptoms. The grading system most commonly used is that described by Fontaine *et al.* in 1954.

Grade 1	No symptoms
Grade IIa	Intermittent claudication >200 metres
Grade IIb	Intermittent claudication <200 metres
Grade III	Rest pain
Grade IV	Ulceration, gangrene

Any patient with a chronic history of intermittent claudication which has not responded to conservative therapy and exercise is a potential candidate for percutaneous transluminal angioplasty (PTA). Patients with clinical Grade IIb usually merit PTA and patients with clinical Grade IIa may also be treated by PTA if their symptoms interfere severely with their work or lifestyle.

Patients with more severe disease, i.e. Grades III and IV, warrant PTA if they have lesions that are technically feasible as this may save the patient from surgery or at least minimize surgery and amputation.

'Blue toe syndrome' is characterized by acute, painful, cyanotic lesions confined to the feet and ankles due to microembolization. Traditionally, surgery has been the treatment of choice, but reports of successful PTA have appeared in the literature (Brewer *et al.*, 1988; Kumpe *et al.*, 1988).

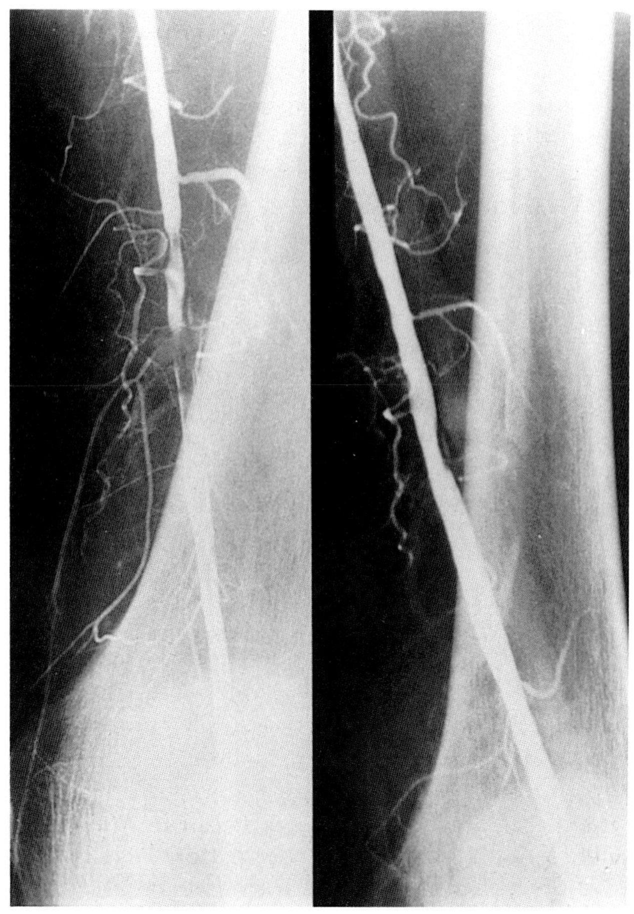

Fig. 1.1 Aspiration thrombectomy
(a) Fresh intraluminal thrombus at the level of the adductor canal. Contrast medium is noted around the thrombus. (b) The lumen is clear of thrombus following aspiration using an 8 Fr wide bore catheter.

Contraindications

1. Acute ischaemic episodes and angiographic evidence of recent thrombosis, i.e. an abrupt occlusion or intraluminal filling defect with poor collateral formation. These should be treated by local fibrinolytic therapy ± aspiration of thrombus and then balloon dilation of any underlying stenosis (Fig. 1.1). (See the chapter on thrombolysis.)
2. Aneurysmal dilation immediately adjacent to a stenosis increases the risk of rupture.
3. Total aortic occlusions.
4. Iliac occlusions >6 cm in length have an increased risk of distal ipsilateral or contralateral embolization (Cumberland, 1982; Ring *et al.*, 1982). The volume of occlusive material can be reduced and the length of occlusion shortened by local thrombolysis, therefore converting a high-risk long iliac occlusion into a shorter one which may be successfully recanalized and balloon dilated with less risk (Auster *et al.*, 1984).

Patient preparation

All patients should be well hydrated. Patients scheduled for morning procedures should receive no food after midnight and should drink clear fluids sufficient only to avoid dehydration.

Patients scheduled for afternoon procedures

should eat a little breakfast (e.g. tea, coffee, toast, cereal) but should eat nothing after this and take clear fluids only.

Routine medication should be taken as normal except that diuretics are usually omitted in order to avoid dehydration and discomfort due to bladder distension in potentially lengthy procedures.

Diabetic patients

Diabetic patients are best scheduled for the morning list if possible. It is helpful if they have an intravenous line placed while still in the ward so that should it become necessary to start an infusion of dextrose this can easily be set up by the nursing staff in the angiography room. Insulin-dependent diabetics should not receive their morning injection of insulin. Noninsulin-dependent diabetics should not receive their morning oral medication. Blood sugars can be monitored during the procedure.

Anticoagulant therapy

Patients on anticoagulant therapy can still be treated by PTA. If on warfarin, the international normalized ratio (INR) should be within the therapeutic range (2–3). Should bleeding be a problem during or post-PTA, the anticoagulant effect can be corrected by infusion of fresh frozen plasma or whole blood. Vitamin K, 5–10 mg, orally or intravenously, can be given to supplement this. This, however, is rarely necessary. All that is required is prolonged pressure at the pucture site until bleeding has stopped.

Patients on heparin are probably easier to control as stopping the heparin before the procedure usually prevents haemorrhage from the puncture site but again, if thrombin time and activated partial thromboplastin times are within therapeutic limits, haemorrhage is unlikely to be a problem. Should haemorrhage occur heparin can be neutralized by protamine sulphate given by slow (10 min) injection. The following guide is recommended by the *ABPI Data Sheet Compendium 1990–91*.

Dose of protamine sulphate (10 mg in 1 ml)	**Time elapsed**
5 ml	Immediately after 5000 IU heparin
2.5 ml	30 min after 5000 IU heparin
1.25 ml	60 min after 5000 IU heparin

Consent

All patients should sign a consent form for PTA with its potential risks and complications fully explained to the patient. This should be done by the radiologist performing the procedure or a deputy. It is important to explain to the patient how the procedure will be performed and what they should expect to feel. The potential complications which might make urgent surgery necessary should be explained and also the fact that the vascular supply to the limb could be made worse by the procedure, also necessitating urgent surgery.

Premedication

Premedication is desirable in most patients, and prescribing practice varies between different hospitals and units. We usually prescribe Omnopon (papaveratum), 10–20 mg, or morphine, 5–10 mg, by intramuscular injection and lorazepam, 1 mg, orally, 1 hour before the procedure. The dose of papaveratum or morphine is varied according to the build of the patient, smaller doses being adequate in small, elderly patients.

Occasionally, when for some reason premedication has not been given before the procedure, we give Diazemuls (diazepam emulsion), 5–10 mg, intravenously.

It is very helpful if patients arrive in the X-ray department for PTA having had both groins, or the relevant groin, shaved in anticipation of the procedure. Many patients are asked to shave their own groins on the ward and are unsure of the precise requirements. Guidance can be given by the radiologist whilst consent is being obtained. This avoids having to shave the patient when in the angiography suite, saving time and pain caused by cleaning a freshly shaved area of skin.

Equipment

Basic requirements

1 Needle
2 Guidewire
3 Guiding and predilating catheters
4 Balloon catheter
5 X-ray equipment with high-resolution image intensifier

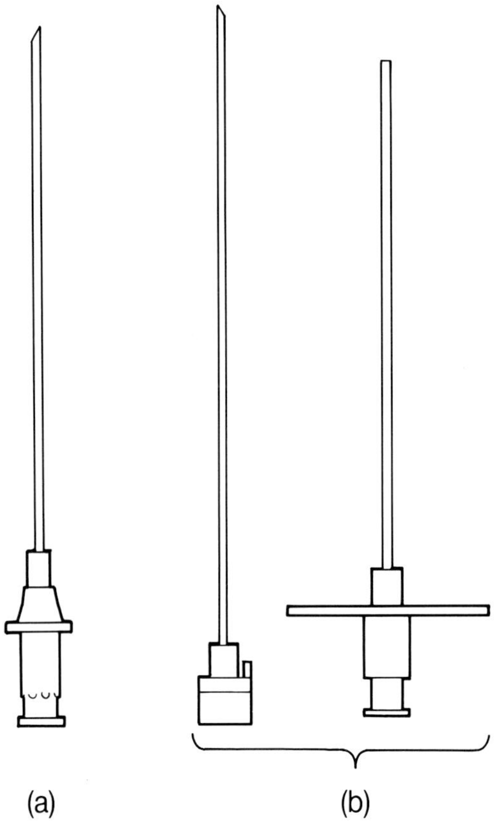

Fig. 1.2 Arterial puncture needles
(a) Single-piece needle. (b) Two-piece needle showing the bevelled needle and blunt inner obturator.

1 Depending on personal preference a single-piece needle or two-piece needle may be used routinely (Fig. 1.2) (see page 12).

2 For angioplasty work, there are a wide variety of guidewires which will help to cross the lesion. In general, a wire with a long floppy tip is useful to start with. Some authors prefer to use a small J tipped wire supported by a catheter to make a passage through occlusions. We prefer a more guided approach using gentle floppy tipped wires, for example, a Newton cerebral or Bentsen wire. A wide angled J wire (7.5 or 15 mm) is useful to try to obtain a different angle. The angled tipped Terumo 'glide wire' has proved a very useful development and resort to this wire is made in all but the easiest lesions. The main disadvantages of this wire are that it has poor tactile feedback and catheters do not readily follow it. We find the angled wire more useful than the straight one as the tip can be torqued whilst tentatively advancing to find a channel through the lesion. Although there are devices available for torque control, simply drying the wire at the catheter hub, with a piece of gauze, works well. To form an angle at the tip of a straight Terumo wire requires tying a knot at the tip and leaving it for several minutes. See Table 1.1 for various useful guidewires and their characteristics.

3 Guiding and predilating catheters are not always necessary. When a guidewire crosses a lesion easily the balloon catheter can be passed over it immediately.

However, for total occlusions and diffuse long segment stenotic disease guiding catheters are necessary both for wire support and for directional control. The van Andel predilating catheter is a strong Teflon catheter with a long tapered tip which gently predilates the lumen through which the wire has passed. However, it is straight and so cannot be used for directing the tip of the wire (Fig. 1.3). For this pupose, a preshaped catheter is necessary. We use femorovisceral catheters as these are usually readily available in the department (Fig. 1.4) (Carver and Cumberland, 1987).

4 A wide variety of balloon catheters are now available, and advances in technology can supply us with strong materials with high burst pressures as well as a low balloon profile.

The choice of balloon catheter is largely a matter of personal preference. There are two types of balloon catheter – the Grüntzig or Olbert. The Grüntzig types of balloon catheters consist of an inner and outer catheter. The outer, enveloping catheter is folded around the inner catheter in an umbrella fashion when deflated. On inflation it unfolds (Fig. 1.5a).

The Olbert types of balloon catheters are more complex. The balloon is firmly attached to the inner catheter distally and the outer catheter proximally. The balloon is stretched in this position so that its diameter is that of the outer catheter and a locking device keeps it in this position (Fig. 1.5b).

The important features are:

(a) The profile of the balloon
(b) The burst pressure of the balloon
(c) The trackability over a wire

(a) The profile of the balloon refers to its cross-sectional diameter and its ability to return to a low profile on deflation compared with the catheter on which it is mounted. A low profile balloon catheter can be pushed over a guidewire through tight lesions with more ease.

(b) Most manufacturers state on the packet the pressure to which the balloon can be safely inflated. Some state the burst pressure. When

Name	Manufacturer*	Characteristics
Newton Cerebral	William Cook Schneider Eschmann	10–15 cm tapered core so floppy tipped and therefore less likely to cause dissection
Bentson	Meadox Schneider William Cook	10 cm tapered core ±5 cm of wire beyond core for extra floppiness
'J'	William Cook Merck Eschman	A variety of 'J' curves are available, ranging from very tight 'J's' (i.e. 1.5 mm diameter) to wide angled 'J's' (i.e. 7.5 or 15 mm diameter)
Glide	Terumo (Radifocus) Meditech	Nitinol core coated with a hydophilic urethane jacket so that the wire becomes very slippery on contact with blood or saline and slides through stenoses with a minimum of friction. The angled tipped wire is particularly useful (see text)
Steerable	Schneider Meditech (PSG) Peripheral Systems Group Target Therapeutics Cordis	Floppy tipped wires with a long flexible, distal part (20–30 cm), stiffer shaft proximally for excellent torque control. Useful in negotiating very tight stenoses particularly in tibial vessels
Stiff	Meditech William Cook PSG	The stiff wires are useful when needing to exchange catheters through very difficult groins (e.g. scarring, obesity) and when it is difficult to advance a catheter over the wire due to very tight proximal stenotic disease

*These are examples used by the authors and not an exhaustive list of suppliers.

Table 1.1 *Types of guidewires useful in percutaneous transluminal angioplasty*

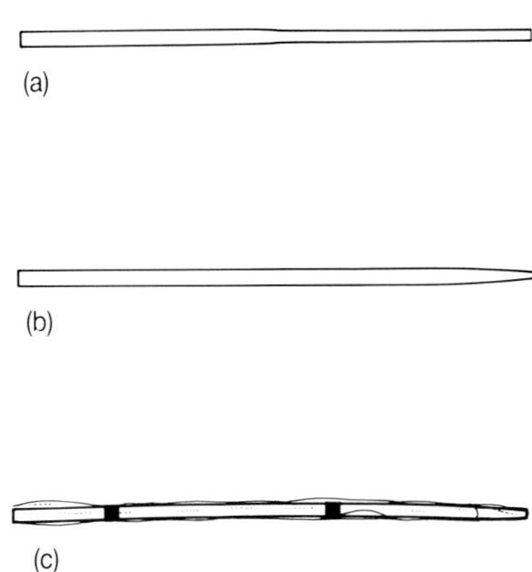

Fig. 1.3 Shape of catheter tips
(a) Van Andel catheter. (b) Vessel dilator. (c) Balloon catheter.

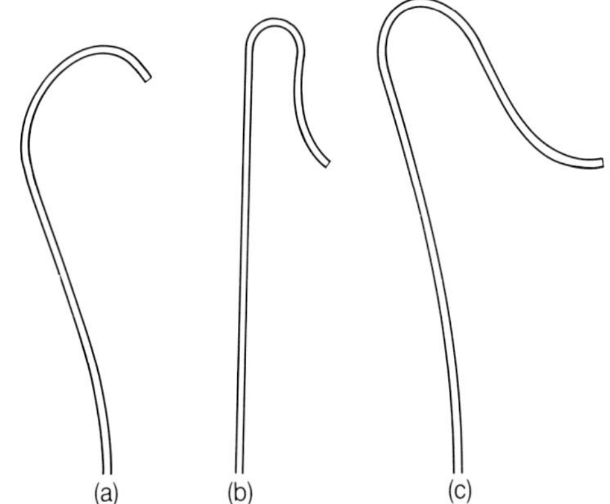

Fig. 1.4 Preshaped catheters
(a) Tip of femorovisceral catheter. (b) Tip of sidewinder 1 catheter. (c) Tip of sidewinder 2 catheter.

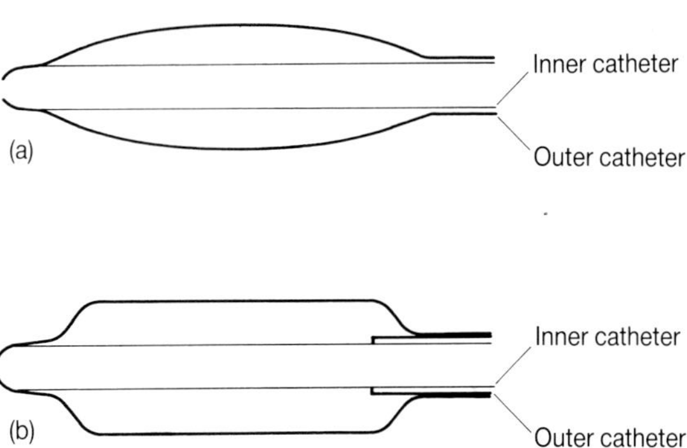

Fig. 1.5 Types of balloon catheter
(a) Grüntzig type of balloon catheter. (b) Olbert type of balloon catheter.

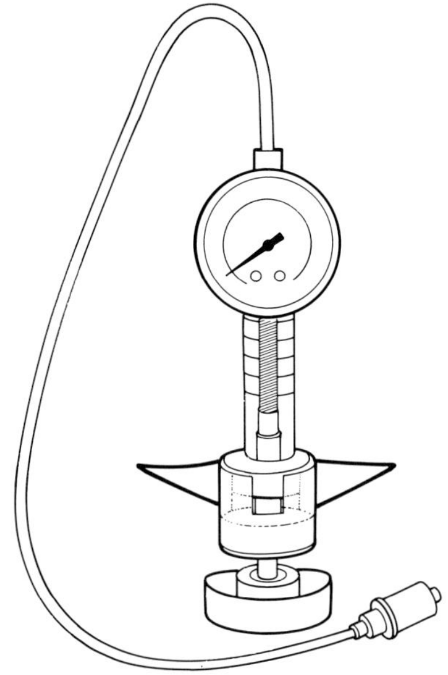

Fig. 1.6 Balloon inflation device
An example of an inflation syringe attached to a pressure gauge, to monitor pressure of inflation.

inflating the balloon, care must be taken generally to avoid higher pressures than those recommended as this may cause the balloon to rupture although there is some leeway. This does not usually cause any problem in the peripheral vasculature but necessitates replacement of the balloon catheter (see the complications section). In tough, fibrous stenoses, for example at graft insertion sites and venous stenoses, a high pressure may be required to obtain a satisfactory angioplasty, in which case high-pressure balloons have to be used. An inflation device with a pressure gauge is helpful so that the pressure in the balloon can be measured, but this is not an essential piece of equipment (Fig. 1.6).

(c) The trackability of a catheter refers to the ease with which it can be passed over a guidewire in very tortuous, curved vessels. Some balloon catheters have a rigid tip which may not easily pass round tight curves, but some of the modern balloon catheters are very soft and flexible and will readily pass over a guidewire in very tortuous vessels.

5 X-ray equipment does not need to be very sophisticated for angioplasty. A high-resolution image intensifier is helpful when dealing with very small vessels. Fluoroscopy with various sized image intensifiers is helpful for peripheral angioplasty work so that magnified views of the area can be obtained, as well as large fields of view for check angiography and a global view of the extent of disease. Oblique views are often necessary and it is much easier to be able to move the X-ray tube and image intensifier than to move the patient. Sophisticated equipment such as for digital subtraction angiography and roadmapping facilities help the radiologist a great deal, and help speed up the time of the procedure.

The procedure may be documented on film, 100 mm camera or ciné, whichever is available.

We do not routinely use catheter introducer sheaths (Fig. 1.7), but if multiple catheter changes are anticipated, or if it difficult to manipulate catheters and wires in the groin, either due to scarring, obesity or puncture of a graft, a sheath may help make the procedure easier for the radiologist and more comfortable for the patient. One size greater than the size of the balloon catheter is usually required, and care must be taken to remember to flush the side arm of the sheath regularly with heparinized saline. We routinely add 1000 IU heparin to 500 ml normal saline on the angiographic trolley.

If a sheath is used, contrast medium can be injected through the side arm to assess the success of PTA below the inguinal ligament, whilst keeping a wire across the site of dilation.

If a sheath is not used, in order to inject contrast medium through the lumen of a catheter and still maintain a wire across the site of dilation, a fine 0.018 inch wire is passed through the balloon catheter and a Tuohy Borst adaptor attached to the catheter hub so that contrast medium can be injected through the side arm whilst the guidewire is in place (Fig. 1.8).

A guidewire should be kept across the lesion at all

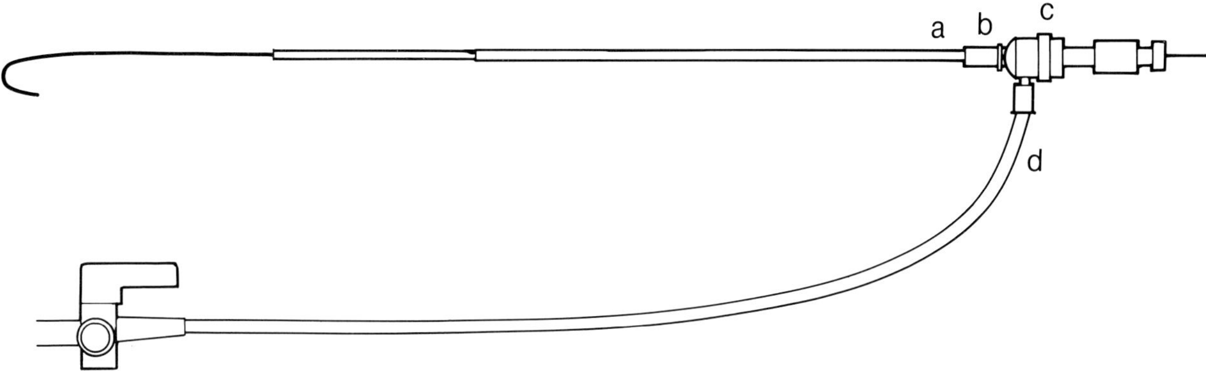

Fig. 1.7 A catheter introducer sheath set (a) Dilator. (b) Sheath. (c) Check-flow valve. (d) Side arm for flushing.

times. Attempts to recross the lumen with the wire after balloon dilation may result in dissection as the endoluminal surface is rough and the wire may easily pick up any intimal tears that balloon dilation creates. If the guidewire accidentally enters one of the tears it may result in dissection and occlusion of the vessel (Fig. 1.9). It is important, therefore, to be able to assess the result of the procedure and whether any further dilation is necessary before withdrawing the guidewire.

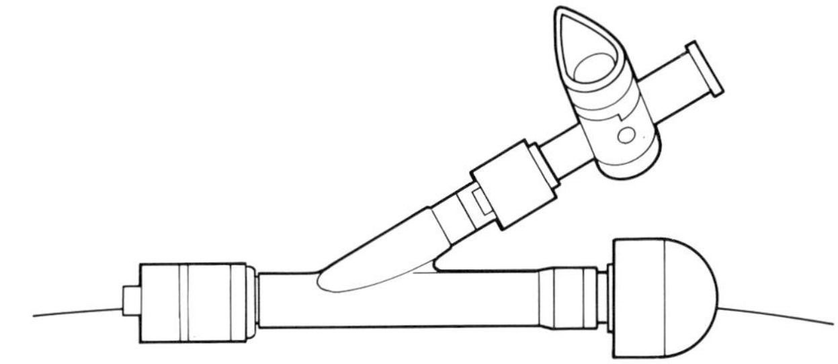

Fig. 1.8 Tuohy Borst adaptor with side port

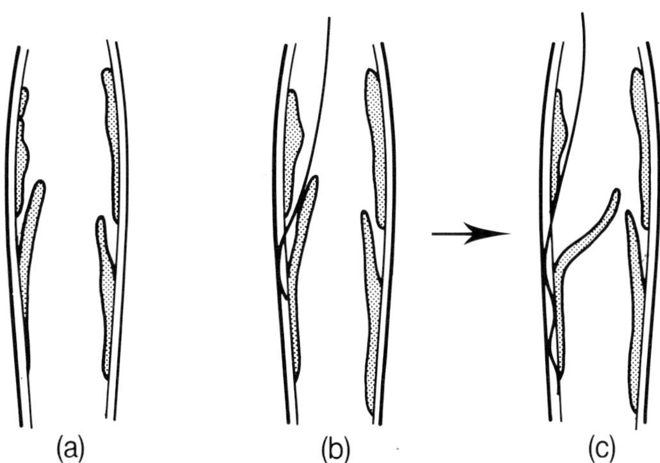

Fig. 1.9 Passing a guidewire through a segment of artery which has undergone balloon dilation
(a) Roughened surface with intimal tears after balloon dilation. (b) Guidewire accidently enters a tear causing dissection. (c) The blood flow further enlarges the dissection channel, leading to occlusion of the lumen.

Arterial access site	Main indications
1 Ipsilateral common femoral artery	All accessible stenotic/occlusive lesions
2 Contralateral common femoral artery	1 Occluded ipsilateral common femoral artery 2 Lesion to be treated is at or very close to the potential puncture site 3 Postsurgical scarring of the ipsilateral groin 4 If early surgery is planned on the ipsilateral limb (to avoid local infection and vascular damage) 5 When two guidewires are required, e.g. if performing PTA at the common iliac or common femoral bifurcation, to protect the arterial branch not being dilated 6 Occasionally when iliac stenoses are difficult to cross from an ipsilateral retrograde approach and have more favourable geometry if approached antegradely
3 Either femoral artery	Aortic and proximal upper limb lesions
4 Low superficial femoral artery	PTA of common femoral artery but contralateral femoral puncture not possible or unlikely to be successful, e.g. common femoral occlusion
5 Low profunda femoris artery	When PTA of a stenosis at the origin of the profunda femoris cannot be approached antegradely
6 Popliteal artery	1 Occlusions involving the origin of the superficial femoral artery 2 Steep, large collaterals at the proximal site of occlusion making antegrade recanalization impossible or if irretrievable dissection has occurred during the antegrade attempt at recanalization
7 Axillary artery	1 Bilateral iliac occlusions 2 Occlusion of the supra-aortic branch involving origin of the vessel. The ipsilateral axillary is punctured 3 Severe aortoiliac disease and vascular tortuosity 4 Downward angulation of vessel to be recanalized, e.g. renal, superior mesenteric artery, so that an approach from the upper limb makes catheterization easier
8 Brachial artery	As above

Table 1.2 *Indications for arterial access site for percutaneous transluminal angioplasty (PTA) of the limbs*

Arterial access

Sites

1. Common femoral, superficial femoral and profunda femoris arteries
2. Axillary artery
3. Brachial artery
4. Popliteal artery

Most procedures can be performed by an ipsilateral or contralateral femoral artery puncture. Indications for various routes are shown in Table 1.2.

Technique

IPSILATERAL FEMORAL ARTERY

For puncture of the femoral artery, the groin must first be shaved and cleaned. The artery is best palpated as it crosses the pubic bone, and as this is also the point at which haemostasis is best achieved this is the best point of access. No regard is made to the inguinal skin crease as this varies in position with obesity and is therefore an unreliable indicator of the position of the inguinal ligament (Grier and Hartnell, 1990). Local anaesthetic (e.g. 2% lignocaine, 5–10 ml) is infiltrated into the tissues all around the artery, including behind the artery. A small skin incision is made with a scalpel blade and a tunnel made through the subcutaneous tissues with a fine artery forceps. This makes catheterization and manipulation easier and helps prevent a haematoma by allowing blood to drain from under the skin. When making the scalpel incision it is important to draw the skin laterally to avoid accidental injury of the underlying artery.

A double or single wall puncture technique can be used safely. It should always be remembered, especially with antegrade punctures, that the angle of entry of the needle into the artery is approximately 45°. This allows smooth entry of catheters through the subcutaneous tissue to minimize buckling of the guidewire, particularly in obese patients (Fig. 1.10). If a sheath is used with a steep angle of entry, the sheath may kink, making it difficult to pass catheters into the artery. This may mean that the site of entry in the skin appears much higher than the actual arterial puncture site. The authors usually use a double wall puncture technique. The advantage of this method is that the bevelled needle is removed leaving a blunt inner obturator within the lumen which can be manipulated from side to side with safety. Such manipulation may be necessary to help the guidewire pass into the superficial femoral artery or profunda femoris artery, whichever is required (Fig. 1.11). The single-piece needle has a sharp, angular point which could damage the internal arterial wall during manipulation (Fig. 1.2). In experienced hands, either method should prove safe and it is then a matter of personal preference. However, there are occasions when a single wall puncture is more desirable (Table 1.3).

Before advancing the guidewire, ensure that there is free pulsatile backflow of arterial blood. If back-

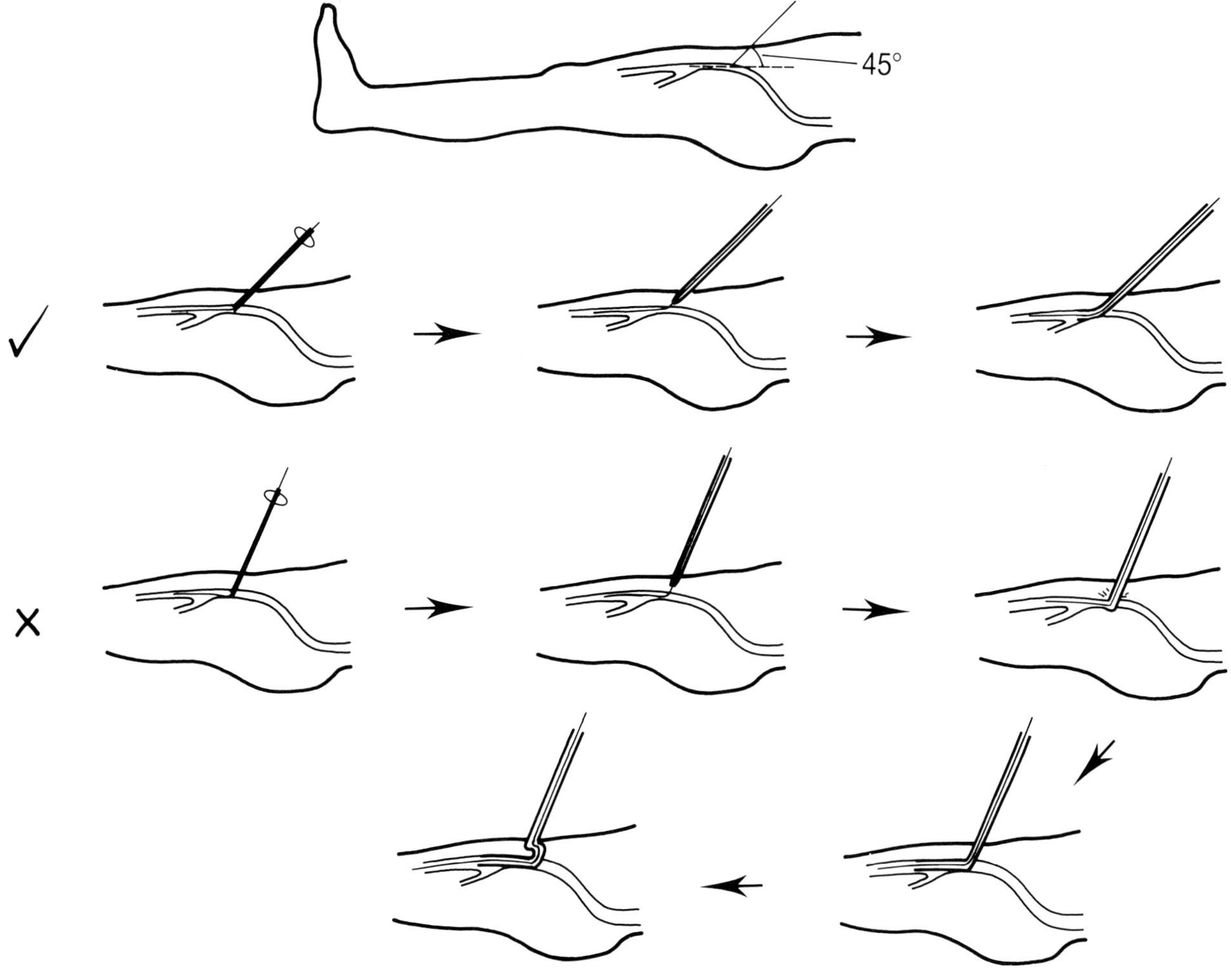

Fig. 1.10 Angle of needle during antegrade femoral artery puncture
(a) Puncture the femoral artery so that the angle of entry of the needle is about 45°. The guidewire is advanced and smooth passage of the catheter over the wire is achieved. (b) A steep angle may make it difficult to advance the wire into the artery and may cause buckling of the wire at the site of entry when attempts are made to advance the catheter. In obese patients this may lead to the catheter forming a loop under the skin.

Indications	Reason
Fibrinolytic therapy	To avoid bleeding from the posterior wall of the artery if systemic lysis occurs
Puncture site in arm or popliteal fossa	The tissues behind the artery are lax so bleeding and haematoma from the posterior wall puncture is more likely
Bleeding dyscrasia	Potential for continued bleeding from posterior wall

Table 1.3 *Indications for single wall puncture*

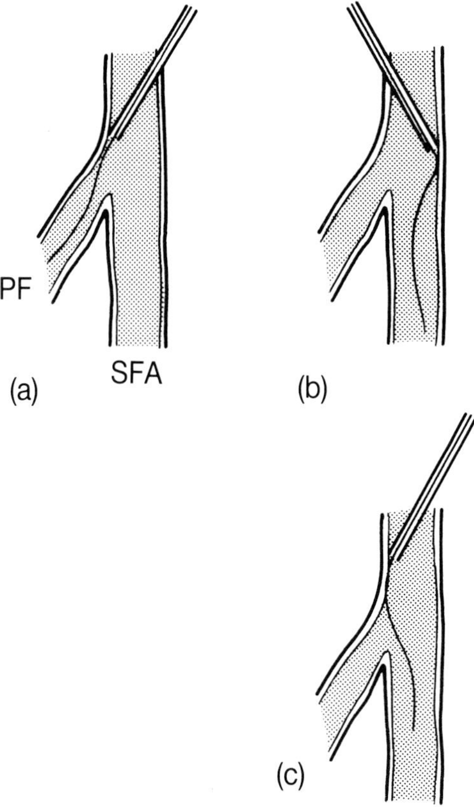

Fig. 1.11 Side-to-side manipulation of the obturator can help the wire pass into the superficial femoral artery (SFA)

(a) The obturator is angled laterally and the wire enters the profunda femoris (PF). (b) The obturator is angled medially and the wire enters the SFA. (c) If the tip of the obturator is away from the origin of the PF, angling the obturator laterally can allow the wire to bounce off the lateral wall of the artery and enter the SFA.

flow is poor, the tip of the needle is probably still partly in the wall, and advancing the guidewire in this situation will result in dissection (Fig. 1.12). If tentative attempts are made to advance the wire at this point, resistance will be felt and the guidewire will not easily advance. If, despite this, the guidewire is pushed forward, the operator will notice a 'sticky' sensation, a tactile sign of dissection and the patient may experience pain. If in doubt the guidewire should be withdrawn and contrast medium injected through the needle. Contrast medium pooling at the site of dissection and along the dissection plane will be noted, and does not flow away on stopping the injection. At this point the needle should be withdrawn slowly until better pulsatile flow is seen or removed completely, haemostasis achieved and the puncture repeated. If, however, the contrast medium flows away freely there is no dissection and the guidewire can be tentatively advanced.

When the guidewire has been advanced sufficiently (i.e. so that there is a good length of wire within the artery to provide an anchor for advancing the catheter), the needle is removed, the wire cleaned with a moist sponge (or wet gauze swab taking care not to leave threads from the swab on the wire) and a dilator advanced over the wire (Fig. 1.3). The dilator is of the same calibre as the catheter to be used, but has a tapered tip. This dilates the artery at the puncture site to the size of the catheter, avoiding damage to the artery wall and blunt catheter tip. If a predilating catheter of the van Andel type is being used, a dilator is not necessary.

For procedures above the inguinal ligament the ipsilateral common femoral artery is the best site of puncture. Retrograde catheterization by the femoral approach is the most familiar puncture method to radiologists and causes few problems. Antegrade ipsilateral common femoral artery puncture is used for infrainguinal lesions. During antegrade femoral artery puncture attention should be paid to the position of the inguinal ligament. The artery should not be punctured above the inguinal ligament as this can result in retroperitoneal haemorrhage which cannot be controlled by manual pressure because of the deep location of the artery and lack of osseous support. The best landmark for antegrade puncture is the centre of the femoral head seen on fluoroscopy. Puncture of the artery at this level should not pass cranial to the inguinal ligament and should in most cases be above the femoral artery bifurcation (Spijkerboer *et al.*, 1990). Review of a previous femoral angiogram of the patient will give you an idea of the best site for puncture. If there is any doubt

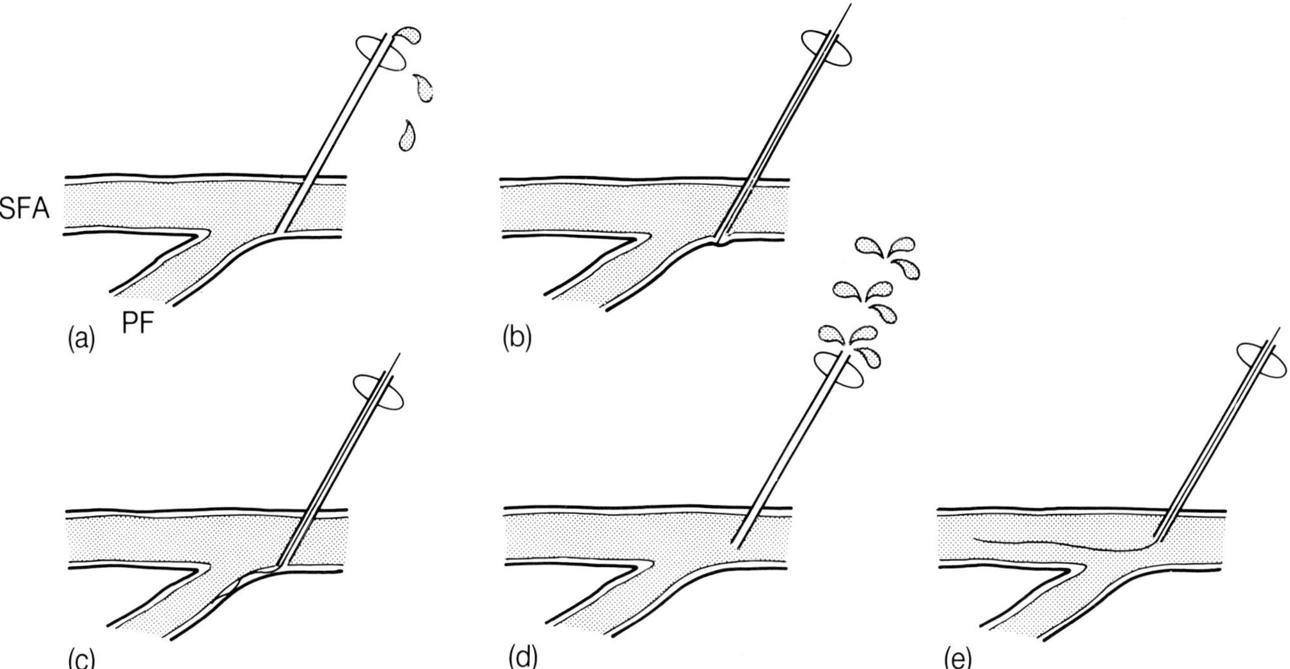

Fig. 1.12 Arterial puncture
(a) Poor backflow indicates the tip of the needle is partly in the wall. (b) The wire does not freely advance. (c) Attempts to advance the wire will result in dissection. (d) Withdrawing the needle frees the tip from the wall and good pulsatile flow is obtained. (e) No resistance is felt when the guidewire is advanced. SFA = superficial femoral artery. PF = profunda femoris.

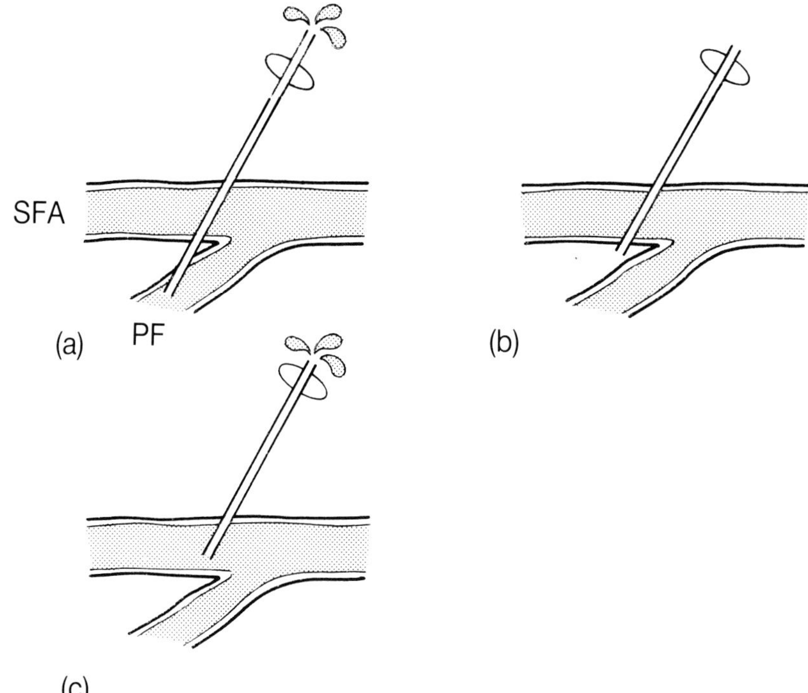

Fig. 1.13 If the profunda femoris (PF) is selectively punctured, withdrawal of the needle may show that the superficial femoral artery (SFA) was punctured 'en route'
(a) The tip of the needle has passed through the SFA into the PF. (b) As the needle is withdrawn, arterial backflow is lost. (c) As the needle is withdrawn further, a second spurt of arterial flow is seen, indicating that the tip is now in the SFA.

it is a straightforward matter to screen the femoral head and assess the distance from the proposed site of puncture on the skin which is marked by a metal pointer or needle.

Puncture of the common femoral artery (CFA) is usually performed for antegrade femoral punctures so that either the profunda femoris (PF) or superficial femoral artery (SFA) can be catheterized. However, on the rare occasions when the femoral artery bifurcates high, selective puncture of the SFA or PF can be performed. If the SFA requires catheterization but the guidewire persists in passing down the PF, an injection of contrast medium will help to define the anatomy and show whether the PF or common femoral artery has been punctured. If the former, it is worth gradually withdrawing the needle to see whether a second spurt of arterial blood is obtained, as occasionally the needle will have passed through the SFA before the PF (Fig. 1.13). If this is the case, catheterization of the SFA is performed without needing a second puncture. Bleeding from the PF in these circumstances does not seem to be a problem.

More often the CFA has been punctured but the guidewire stubbornly persists in selecting the PF. Various manoeuvres can help guide the wire into the SFA. Simple external rotation of the ipsilateral limb can be helpful. Flattening the needle against the patient's abdomen elevates the tip of the needle helping selection of the SFA (Fig. 1.14). If the SFA is still not selected movement of the needle cannula from side to side can be helpful (Fig. 1.11). Occasionally despite these manoeuvres, the guidewire still fails to enter the SFA. If this proves to be the case, a preshaped catheter, for example, a femorovisceral/cobra, can be passed into the PF (Fig. 1.15) and slowly withdrawn while injecting contrast medium until the origin of the SFA is identified. The catheter is then turned so that the tip of the catheter points down the SFA and a guidewire is then passed.

If the PF and SFA are projected over each other, it

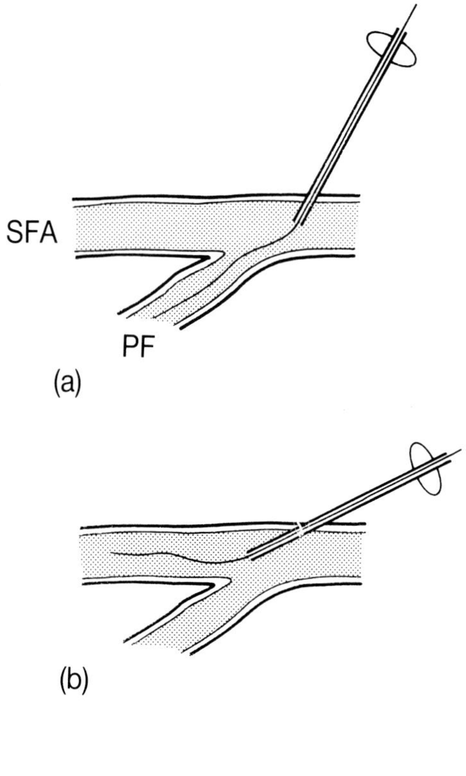

Fig. 1.14 Flattening the needle against the abdomen elevates the needle tip helping the wire to pass selectively into the superficial femoral artery (SFA)
(a) Guidewire repeatedly enters the profunda femoris.
(b) Flattening the needle lifts the tip and the guidewire selects the SFA.

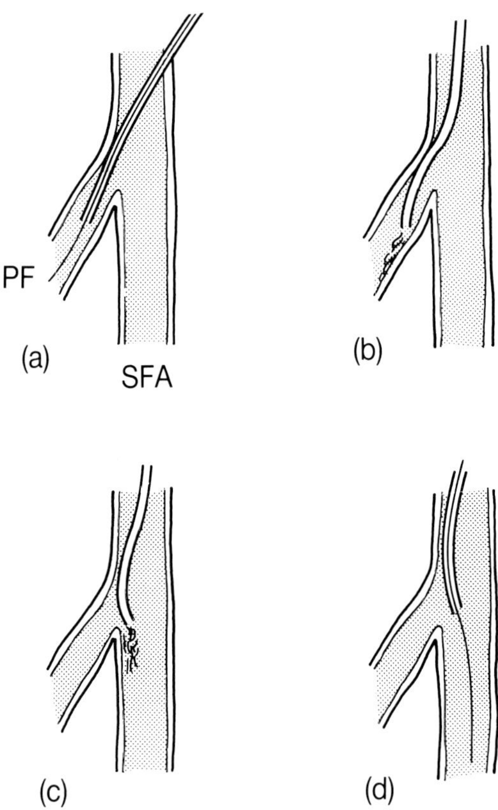

Fig. 1.15 A preshaped curved catheter, e.g. femorovisceral can be used to guide the wire into the superficial femoral artery (SFA) when the profunda femoris (PF) is persistently catheterized after common femoral artery puncture.
(a) The catheter is advanced over the wire in the PF. (b) The wire is withdrawn and contrast medium injected as the catheter is slowly withdrawn. (c) The origin of the SFA is identified and the catheter turned so that it points to the SFA. (d) A guidewire is advanced into the SFA.

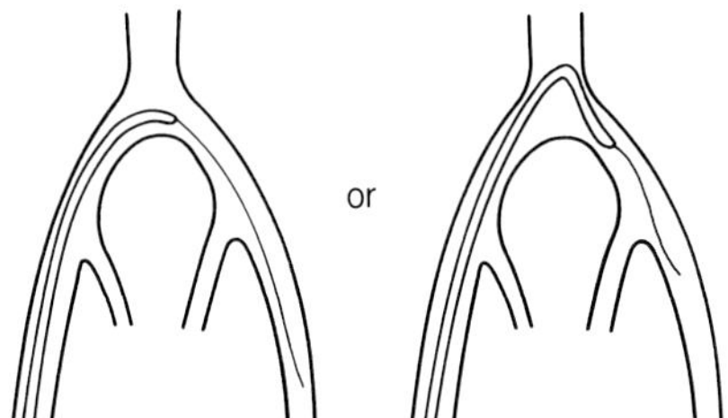

Fig. 1.16 A sidewinder or femorovisceral catheter is used to cross the aortic bifurcation, taking care not to dissect the common iliac artery.

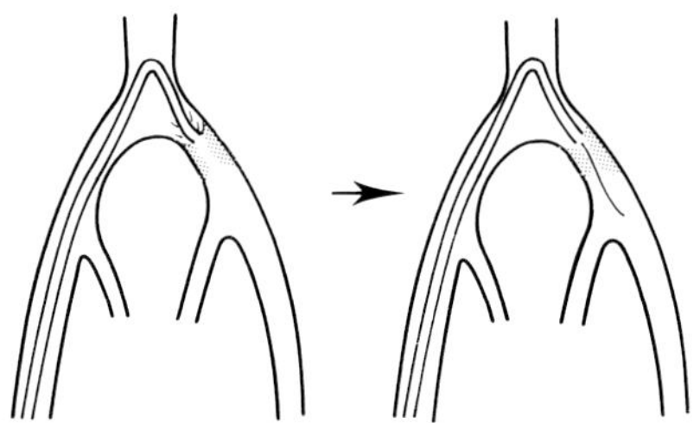

Fig. 1.18 If the contralateral lesion is close to the bifurcation the tip of the catheter may cause dissection. This can be avoided if a floppy tipped wire projects a few centimetres from the tip of the catheter.

can be difficult to know when the SFA has been selected and an oblique projection helps clarify the situation.

CONTRALATERAL FEMORAL ARTERY

The technique is the same as for ipsilateral femoral catheterization, the difference being that the catheter and wire have to pass over the aortic bifurcation. Usually this can be easily achieved by either femorovisceral or sidewinder catheters (Fig. 1.16). Usually a sidewinder 1 is all that is necessary. If the occlusion/stenosis to be treated is well away from the origin of the common iliac there should be no difficulty (Fig. 1.17), however, if the lesion involves or is close to the bifurcation care has to be taken particularly when using a sidewinder so that the limb of the catheter does not dissect the arterial wall (Fig. 1.18). Allowing the tip of a floppy tipped guidewire to protrude a couple of centimetres may help guide the sidewinder catheter through the stenosis. The indications for a contralateral femoral approach are shown in Table 1.2.

Problems which may be encountered on using a contralateral femoral approach include negotiation of the balloon catheter over an acute aortic bifurcation. This is usually overcome by using a stiff wire, for example, a Rosen wire or an Amplatz superstiff wire. Most catheters will follow these stiffer wires especially if the tip of the wire is passed well down the iliac artery to serve as an anchor, and many of the 5 Fr catheters are particularly good at tracking over the wire (Fig. 1.19). In chronic occlusions inadequate forward pressure may be a problem in the contralateral approach as the wire has a tendency to buckle into the abdominal aorta, and catheter manipulation is made particularly difficult if the iliac arteries are very tortuous and irregular.

Puncture of a femoral artery when a pulse cannot be palpated is perfectly feasible. Patency of the iliac and common femoral artery should first be confirmed by either angiography (intra-arterial angiography or intravenous digital subtraction angiography) or by duplex ultrasound. Ultrasound or bony landmarks may be used to locate the common femoral artery which can often be palpated

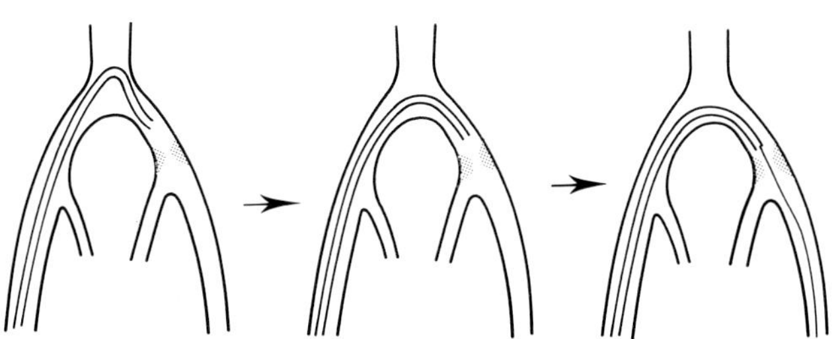

Fig. 1.17 As the sidewinder catheter is pulled down over the aortic bifurcation, the tip remains clear of the contralateral lesion, and a guidewire can then be advanced safely through the lesion.

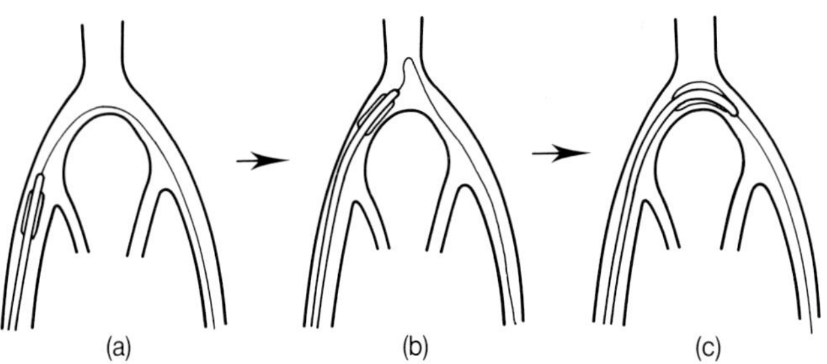

Fig. 1.19 A stiff wire may help a balloon catheter pass over the aortic bifurcation. (a) A guidewire is passed well down the contralateral iliac artery. (b) If the guidewire is not stiff, the balloon may cause the wire to buckle at the bifurcation and not pass over it. (c) If a stiff wire is used, the balloon catheter passes more easily.

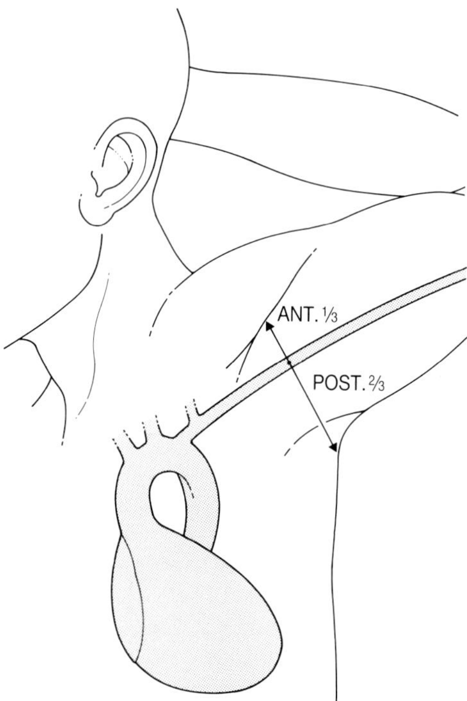

Fig. 1.20 Patient and arm position for axillary artery puncture.

even though it is pulseless. Fluoroscopy of the femoral head can be used to direct puncture over the medial aspect of the femoral head or identify calcification in the vessel wall and so direct puncture. Using any combination of these methods successful puncture of a patent femoral artery should be accomplished.

AXILLARY ARTERY

When angiography and PTA are not possible from the femoral artery approach or if the direction of the vessel to be treated is very acute from a femoral approach, an axillary or brachial approach can be used.

The main indications are shown in Table 1.2. To puncture the axillary artery the patient should be lying supine with the hand positioned behind the head so that the arm is at right angles to the trunk (Fig. 1.20). In this position the artery is nearly straight and can be palpated against the humerus at the junction of the anterior with the middle third of the space between the anterior and posterior folds of the axilla. The axillary artery is very closely related to the brachial plexus and at the site of puncture the median nerve lies on its outer side and the ulnar nerve on its inner side, separating the artery from vein. It is important to avoid injury to these structures by fibrosis following haematoma formation. For this reason a single wall puncture technique is preferred using a single-piece needle. It is also advisable to use a small catheter (e.g. 5 Fr) as the artery is relatively small. The left axillary artery is usually punctured as this allows a more direct route to the abdominal aorta. One of the disadvantages of an axillary approach is the distance from the arm to the lower limbs. This makes it necessary to use long wires and catheters and because of the distance, and lack of forward pressure and directional control, it may be more difficult to advance the angioplasty catheter through the lesion. Control is compromised by the angle the catheter makes as it passes from the subclavian artery into the aorta. Tortuosity and atheromatous disease of the subclavian artery may also contribute to poor directional control. For these reasons, an axillary or brachial artery approach to lesions below the inguinal ligaments is rarely useful.

BRACHIAL ARTERY

Percutaneous puncture of the brachial artery may be performed midway between the axilla and antecubital fossa (a high brachial puncture) or at the antecubital fossa. Many authors would advise a surgical cutdown onto the brachial artery in the antecubital fossa rather than percutaneous puncture

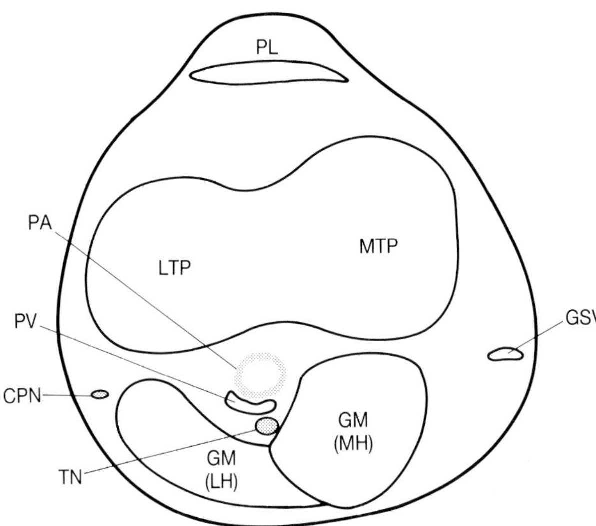

Fig. 1.21 Cross-section of the popliteal fossa showing the relationship of the vein and nerve to the popliteal artery.
PL = patellar ligament. MTP = medial tibial plateau. LTP = lateral tibial plateau. GM(MH) = gastrocnemius muscle – medial head. GM(LH) = gastrocnemius muscle – lateral head. PA = popliteal artery. PV = popliteal vein. TN = tibial nerve. CPN = common peroneal nerve. GSV = great saphenous vein.

but such a technique requires experience (Sones and Shirley, 1962; Fergusson and Kamada, 1981). A high brachial puncture has been shown to be safe and effective (Gaines and Reidy, 1986; Baudouin *et al.*, 1990), but like puncture of the axillary artery, a single-piece needle should be used and the smallest possible catheter to avoid haematoma and spasm of the artery. The brachial artery at this point is not contained within a sheath and so haematoma formation is unlikely to cause compression of the brachial artery or cause permanent damage to the median nerve.

POPLITEAL ARTERY

Puncture of the popliteal artery is increasing in popularity. The indications are described in Table 1.2. The patient lies prone and ultrasound guidance may be used to direct puncture of the popliteal artery. The smart needle (Peripheral Systems Group) is a single piece needle with a 14.3MHz piezoelectric crystal which provides a continuous Doppler signal to aid puncture of the artery. One of the main theoretical disadvantages of the popliteal approach is that the popliteal vein usually lies immediately posterior to the artery in the popliteal fossa and avoidance of this structure may not be possible (Fig. 1.21). The skin incision is usually made just above the level of the knee joint. The smart needle helps direct the puncture towards the artery and avoiding the vein by allowing differentiation between the doppler signals. In an effort to avoid haematoma formation, it is advisable to refrain from administering heparin until haemostasis has been achieved.

OTHER SITES

Low puncture of the superficial femoral artery or profunda femoris in the thigh may occasionally be required, particularly when treating lesions close to the femoral bifurcation or in the common femoral artery when a contralateral femoral approach cannot be used. The method of puncturing these vessels can be the same as the method of puncturing the popliteal artery, but without needing to turn the patient prone. The main problem with these sites is the loose surrounding tissue making it difficult to compress the artery adequately following the procedure.

BY-PASS GRAFTS

There is no real contraindication to puncturing grafts for PTA. Vein grafts are no more difficult to puncture than native vessels, but prosthetic grafts provoke a dense fibrous reaction, making catheterization difficult. In these circumstances a stiff supporting guidewire may be necessary and a vascular access sheath will help reduce friction between the catheter and surrounding tissue. Occasionally when dense scarring prevents passage of a catheter, dilating the tract briefly to one French size greater than the catheter may help. Pericatheter bleeding can be a problem, though, if this is overdone. The main risk of puncturing grafts is infection at the puncture site which may compromise the viability of the graft, and this may be avoided by prophylactic antibiotic cover.

PEROPERATIVE ARTERIAL PUNCTURE

If the artery is to be surgically exposed, entry can be made either by a direct single wall needle puncture or by arteriotomy. Needle puncture has the advantage of the catheter sealing the puncture site without decreasing the distal flow. It has the added advantage of familiarity for the radiologist, there being a tendency for loss of 'feel' when manipulating through an open arteriotomy.

Recanalization technique

There are variations to the technique depending on the operator's personal preference and experience. The methods described here are those used by the authors.

Once arterial access has been achieved, the main objective of the operator is to cross the stenosis or occlusion with a guidewire. The wire should be gently negotiated through the lesion and there is no place for brute force which may result in lifting an intimal flap and completely occluding the lumen. The technique varies slightly for stenoses and occlusions.

Stenosis

A floppy tipped guidewire should be used (e.g. Bentson, Newton-cerebral, Terumo) with a straight catheter in the first instance. The catheter acts as a support to make negotiation of the obstructing lesion easier. The catheter is advanced to within a few centimetres of the tip of the wire. The wire is then slowly advanced, manipulating it gently so that it does not buckle but passes through the lumen. At this point wires with differing curves can be used to help find the arterial lumen, or a preshaped catheter, for example, a cobra, can be used instead of a straight catheter to help the wire probe the lesion in different directions (Fig. 1.22). The guidewire should always advance before the catheter. If the catheter leads dissection is more likely to occur. When the lesion has been traversed, the catheter is passed over it to predilate the lumen.

If there are multiple stenoses over a long segment the guidewire is advanced as far as it will go easily and if it buckles or a resistance is felt, the wire is kept back at a point before the resistance and the catheter advanced behind it to within a few centimetres of the tip of the wire. The wire is then gently advanced

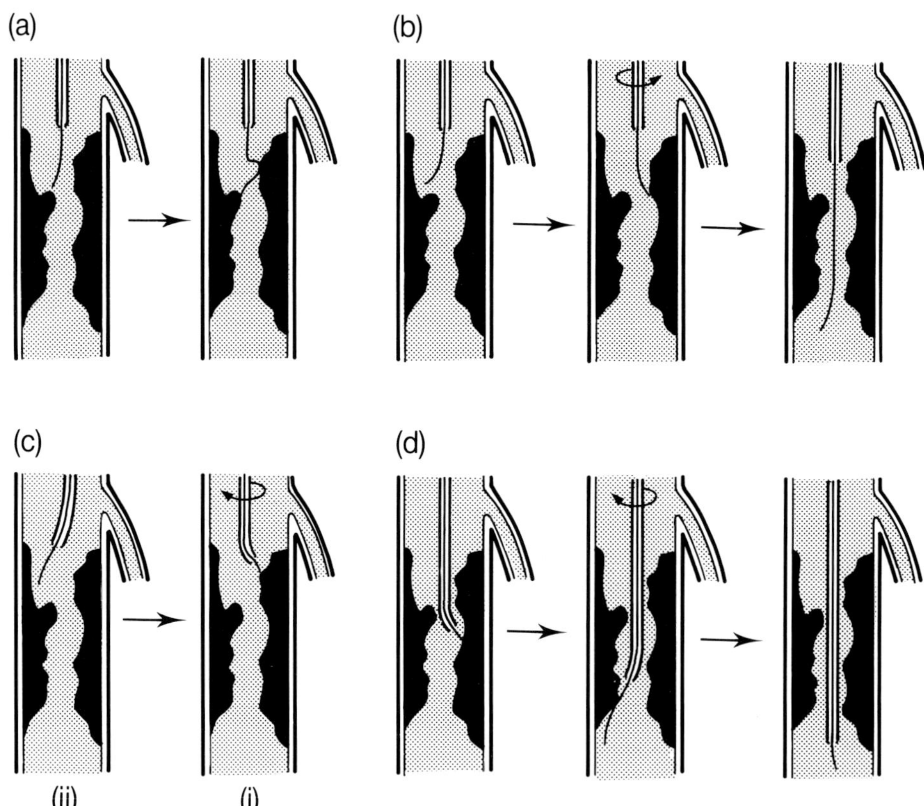

Fig. 1.22 Steps in crossing a stenosis
(a) A straight catheter and wire impinge on plaque. (b) A curved wire is used and is rotated to point in different directions in order to find a path through the stenosis. (c) A curved catheter is used to direct the wire away from the plaque. (d) As the catheter is advanced over the wire, it can still be rotated to guide the wire through the stenosis.

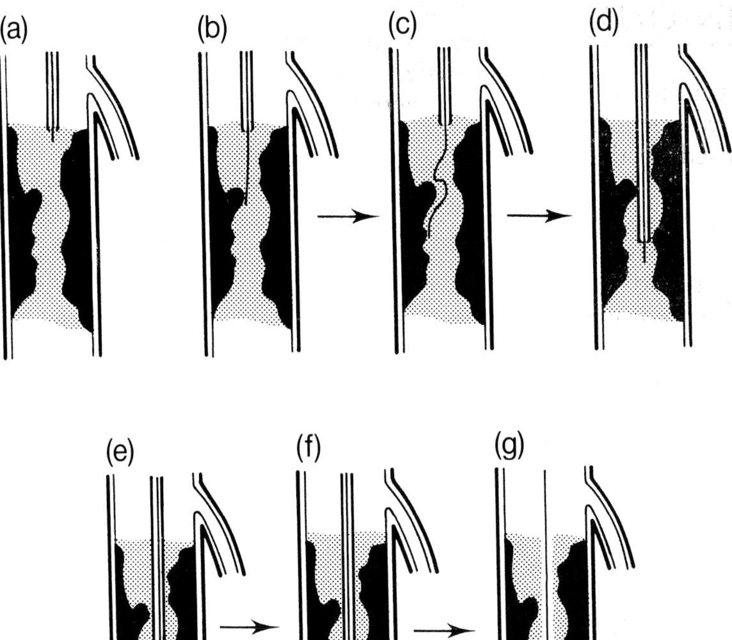

Fig. 1.23 Steps in crossing an occlusion
(a) Catheter is advanced near to the tip of the wire. (b) The wire is advanced through the occlusion. (c) The wire buckles as it impinges on underlying atheroma. (d) The catheter is advanced over the wire up to the point of resistance. (e) The wire is advanced further when supported by the catheter. (f) The catheter is passed over the wire into distal patent lumen. (g) This predilates a channel through the occlusion.

further by probing in different directions until it can be easily advanced. The catheter is again advanced over it until the diseased segment is completely traversed in this step-by-step method (Fig. 1.22).

Complete occlusions

The same method is used but the catheter is advanced to within a centimetre of the tip to give extra support (Fig. 1.23). A curved catheter may be used in an effort to guide the wire away from an important collateral at the proximal end of the occlusion (Fig. 1.24). If the wire accidentally passes subintimally, this can be recognized by a 'sticky' sensation and spiralling of the guidewire, with an impression of being in a wider space than the arterial lumen (Fig. 1.25).

If dissection occurs, any injection of contrast medium will stay near the catheter tip, whilst injection into an occlusion usually fills collaterals and may help delineate the path to the nonoccluded segment. Should dissection occur, the catheter and wire are withdrawn to a point when an intraluminal position is definite and a new path is sought by pointing the wire in a direction away from the dissection

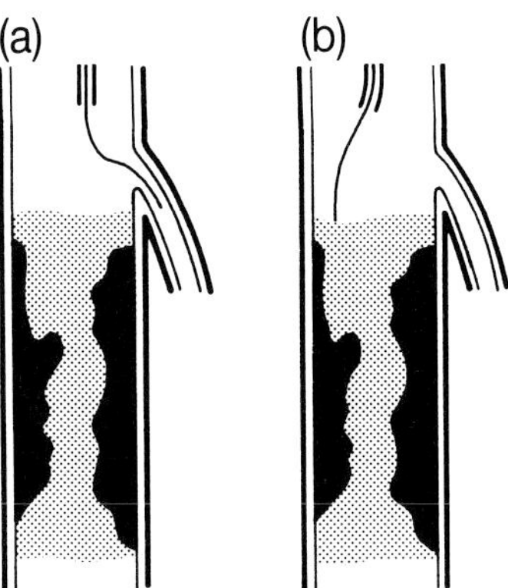

Fig. 1.24 A curved catheter helps to direct the guidewire away from important collateral vessels

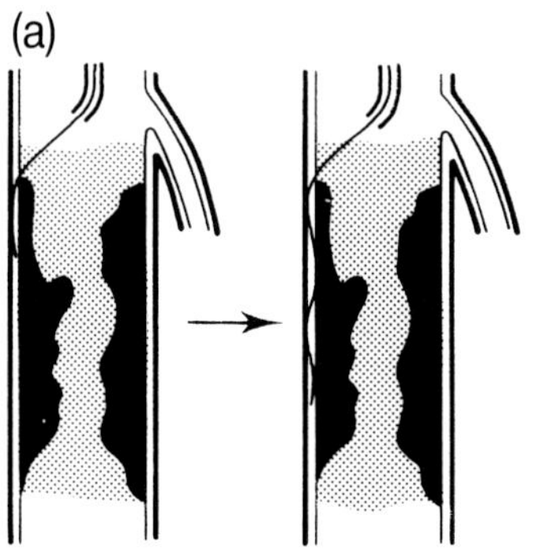

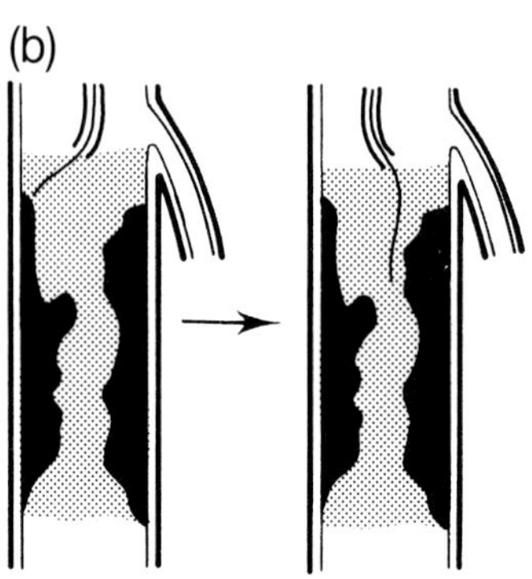

Fig. 1.25 Guidewire dissection
(a) The wire passes subintimally. This can be recognized by tactile feedback and the curving appearance of the wire in the subintimal space. (b) The catheter is used to direct the wire away from the dissection.

(Fig. 1.25). An oblique projection or the use of a curved catheter can help direct the wire away from the dissection in an effort to find a new plane. If the wire re-enters the dissection plane, no resistance is felt. If the wire picks up a different channel there will be a sensation of some resistance. The wire is then advanced tentatively with the catheter close behind to give support, and step-by-step progress is made until the wire is found to be free in the distal arterial lumen.

Balloon dilation

The balloon may have to be inflated in different positions to cover the whole length of the lesion. For long lesions, a long balloon will help minimize the number of repeat inflations necessary. The diameter of the balloon should correspond to the diameter of the arterial lumen measured in a reasonably healthy, nonstenosed segment, just above or below the lesion. Direct measurements can be made in conventional film angiography. If using digital subtraction angiography (DSA) or 100 mm film, a rule placed between the legs can be helpful in calculating diameter, as this can be used for calibration on the DSA console. In practice, no sophisticated technique is necessary, and a visual estimate of the diameter will usually suffice.

Rough guide for balloon diameter

Iliac segment	8–10 mm
Superficial femoral artery	5–6 mm
Popliteal artery	4–5 mm
Tibial arteries	2–3 mm

If the balloon diameter is too large for the artery, the patient will usually complain of unbearable pain as the vessel is dilated in which case a smaller diameter balloon should be used. If the balloon diameter is appropriate, the patient may feel a dull ache or tightness which they can just bear. Severe pain indicates overdilation and risks arterial rupture. Dilation is performed by inflating the balloon with a 1:1 contrast medium:saline mixture and inflation pressures of 5–8 atm (500–800 kPa). Various infla-

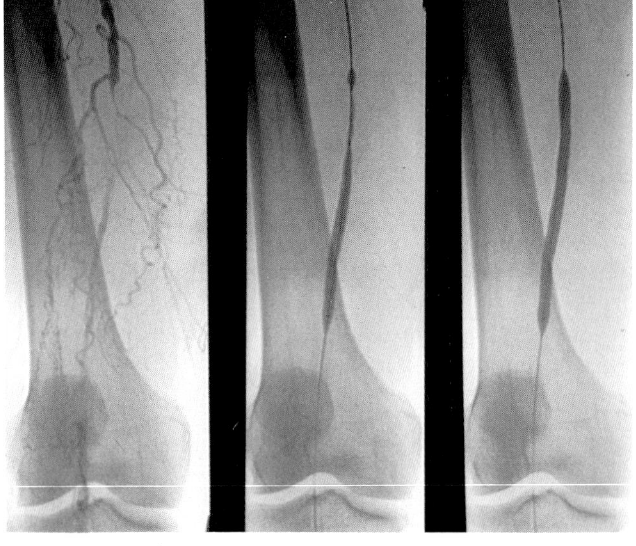

Fig. 1.26
(a) Femoropopliteal occlusion. (b) The balloon is not fully inflated, with waisting proximally at site of occlusion. (c) Full inflation of the balloon.

tion devices are available, some of which include a pressure gauge (Fig. 1.6). There is no consensus as to the correct duration or number of inflations. The balloon may be inflated for 10–15 sec or up to 1 min, and repeated if there is still an unsatisfactory lumen. The most important point is that the balloon inflates fully and there is no residual waisting on the balloon (Fig. 1.26). If waisting cannot be abolished, a balloon which can be inflated to higher pressures may be required or a slightly larger balloon, for example, of 6 mm instead of 5 mm diameter.

Occasionally, difficulty is encountered in advancing the balloon catheter over the guidewire due to very tight stenotic disease proximally. Passing a predilating catheter first (e.g. Van Andel type) may help passage of the balloon catheter. If not, balloon dilation at the site at which the balloon catheter passes no further usually facilitates distal passage of the catheter. If this also fails an extra stiff guidewire (but with a floppy tip) may have to be passed through the catheter.

Has adequate balloon dilation been performed?

AORTOILIAC SEGMENT

Emphasis is placed on the abolition of a systolic pressure gradient across the treated segment. Systolic pressure measurements are made proximal and distal to the lesion before and after PTA. Measurements may be made at rest, but if there is a gradient of ≤10 mm Hg it is worth checking whether the gradient increases after stressing with a vasodilator such as papaverine, 30 mg, injected over 1–2 min. If the pressure gradient increases it is then worth going on to balloon dilation; if not, the lesion is unlikely to be significant. A resting gradient of ≤15 mm Hg is considered significant. After PTA the optimum result is that no residual pressure gradient exists across the lesion (Fig. 1.27). Repeated balloon dilation is performed if the gradient persists, but a gradient of ≤10 mm Hg or less than 50% of the initial predilation value is accepted (Belli *et al.*, 1989). If a gradient of >15 mm Hg remains in the iliac artery a case for stenting may exist.

Pressure measurements are made at the tip of the catheter by attaching a pressure transducer to the catheter via a Y-connector or Tuohy Borst connector. This is possible whilst keeping a fine (0.018 inch) guidewire across the lesion (Fig. 1.8).

An angiogram following PTA can be performed for documentation but the angiographic appearance is a poor indicator of success, as intimal dissection often reduces the aesthetic quality of the angiogram.

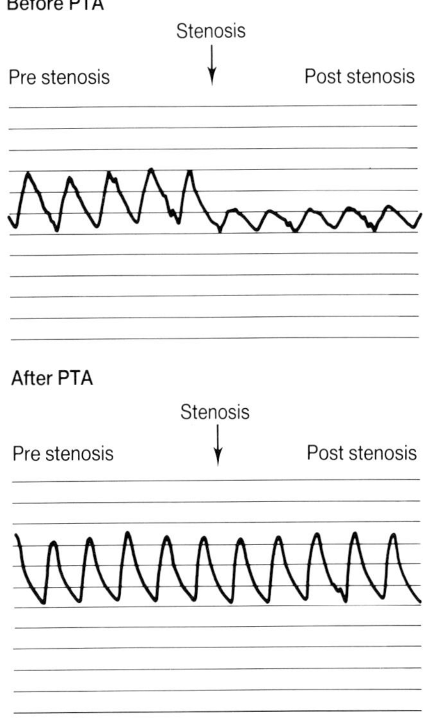

Fig. 1.27 A 40 mm Hg systolic pressure gradient across an aortoiliac stenosis is abolished by balloon dilation. PTA = percutaneous transluminal angioplasty.

BELOW THE INGUINAL LIGAMENT

Pressure readings are less accurate below the inguinal ligament and more emphasis is placed on the angiographic appearance before and after PTA. The aim of angioplasty is to abolish any residual stenosis as a residual stenosis of ≥30% is associated with a higher degree of recurrence.

Subclavian percutaneous transluminal angioplasty

The technique varies little from the standard technique of balloon dilation described for the lower limb. It is usually performed for symptoms of arm claudication or subclavian steal syndrome. Selective catheterization of the subclavian artery is required. After crossing the lesion the preshaped diagnostic catheter is exchanged for a balloon catheter of appropriate diameter (usually 8 mm). A long exchange wire is usually necessary to allow this exchange of catheters without losing the guidewire position. The main concern is the proximity of the vertebral artery to most subclavian stenoses, but it has been shown that it is safe to inflate the balloon across the origin of the vertebral artery if the vertebral artery appears to arise from a relatively normal segment of artery or one only involved by poststenotic dilation (Vitek, 1989). Measurement of pressure gradients across the lesion can be used to assess whether adequate balloon dilation has been performed in the same way as for iliac lesions.

Percutaneous transluminal angiography of renal dialysis fistulae

The stenoses are often on the venous side of the fistula and because of their fibrous nature, long dilation times of 1–2 min and repeated inflations (up to 10) may be necessary. High-pressure balloons may be helpful. Some authors use balloons that are 'oversized' but most authors use balloon diameters of 6–8 mm. It is very important on the initial angiogram to assess the patency of more central veins, as proximal stenoses develop in up to a quarter of patients (Glanz *et al.*, 1987).

Dilation in Cimino and graft fistulae is very painful and local infiltration with anaesthetic may be helpful. The result of balloon dilation is usually assessed angiographically (Fig. 1.28).

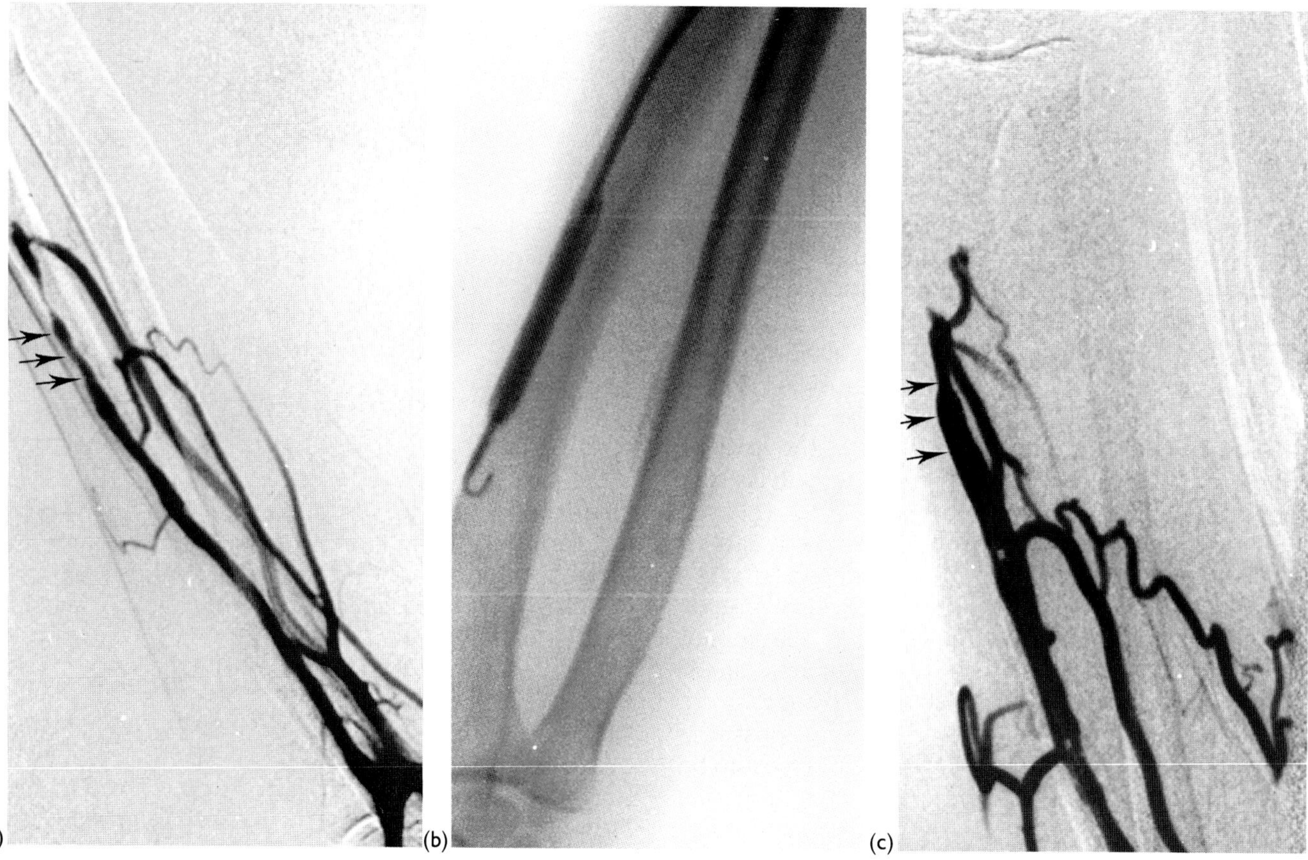

Fig. 1.28 Balloon dilation of a stenosis on the venous side of a dialysis fistula
(a) Before percutaneous transluminal angioplasty (PTA). (b) During balloon inflation. (c) After PTA.

Postprocedural care

When the procedure is complete and haemostasis at the puncture site has been achieved, the patient is returned to the ward. Twenty-four hours' bed rest is advised. The patient must lie absolutely flat in bed for the first 3 hours, and is then allowed to be elevated slightly with pillows behind the head. Pressure over the puncture site during coughing, straining or raising the head is advised and the patient is shown how to do this while still in the angiography suite.

The patient should be kept on clear fluids only until it is clear that urgent surgical intervention is not warranted, i.e. 1–2 hours postprocedure.

Blood pressure measurements, pulse rate and observation of the arterial puncture site are advised at regular intervals to ensure no bleeding is occurring. The following regime is recommended:

Observation every	15 min for	1 hour
Observation every	30 min for	1 hour
Observation every	60 min for	2 hours
Observation every	2 hours for	2 hours
Observation every	4 hours for remainder of period	

The following day, after 24 hours' bed rest the patient is allowed to mobilize. With the increasing use of 5 Fr balloon catheters, 12 hours' bed rest may suffice but too early mobilization will increase the risk of haematoma formation. This has meant that PTA can be performed on an outpatient basis and this is already happening in some institutions.

We do not advocate the use of heparin following PTA unless the procedure has been particularly complicated and an increased risk of thrombosis is suspected, for example, long occlusions and diffuse atherosclerotic disease. In such cases the patient should be fully heparinized, which will delay the patient's discharge.

If there are no complications, the patient can be discharged after 24 hours. It is recommended that the patient is accompanied home by a relative or friend. Vigorous or strenuous exercise should be avoided for the first week but walking is to be encouraged. The general practitioner should be informed of the patient's discharge and treatment as soon as possible. An appointment should be made for the patient to be reviewed in the clinic in 1 month. At this time advice on the importance of regular exercise, nonsmoking and possibly diet can be reinforced. We usually recommend that the patient continues on low-dose aspirin, 75–150 mg, daily. As many patients with peripheral vascular disease have disease elsewhere, there is a good case for them staying on aspirin indefinitely if tolerated. If aspirin is contraindicated, dipyridamole, 50–100 mg tds, may be instituted.

Complications and their management

Complications of arterial puncture

Incidence of complications requiring treatment or complicating patient care vary according to site of puncture (Hessel *et al.*, 1981):

Femoral artery	1.7%
Translumbar aortography	2.9%
Axillary artery	3.3%

The types of complication that may occur at the site of arterial puncture are shown in Table 1.4.

Haematoma
Arterial occlusion/thrombosis
Arteriovenous fistula
Pseudoaneurysm
Neural damage
Abscess
Impaction of balloon catheter

Table 1.4 *Complications at the puncture site*

Complications at the puncture site

HAEMATOMA

Firm digital pressure is applied following removal of the arterial catheter, sufficient to prevent a haematoma but not to occlude distal arterial flow. Control is much better effected by hand control rather than by any commercially available compression devices. Pressure should be applied for as long as it takes to stop bleeding, a minimum of 10 min. If bleeding continues and the patient has been heparinized, protamine sulphate may be given to reverse the effects of heparin (see earlier section). Following angioplasty, 24 hours' bed rest is advised as earlier mobilization increases the incidence of haematoma. Most haematomas only require conservative therapy, but occasionally surgical evacuation and/or repair of the underlying artery is necessary. In a series of patients studied by the authors haematoma formation requiring surgical intervention occurred in 0.12% (Belli *et al.*, 1990a). Other authors have reported an incidence of haematoma of 4%, but only 0.28% required surgical intervention (Gardiner *et al.*, 1986). Damage to the artery may occur if the balloon fails to deflate adequately. Such damage is less likely to occur if an introducer sheath has been used and the balloon can be pulled back into the sheath. If the balloon completely fails to deflate (which fortunately is very rare), percutaneous puncture and aspiration may be necessary. Difficulty in withdrawing a balloon catheter where the balloon has ruptured is a rare problem as balloons are designed to rupture longitudinally. If the balloon cannot be withdrawn percutaneously (e.g. if it has been avulsed) surgical cutdown may be necessary.

ARTERIOVENOUS FISTULA, PSEUDOANEURYSM AND THROMBOSIS

These are consequences of damage to the artery at the puncture site. Pseudoaneurysm and thrombosis at the puncture site occur with a reported incidence of between 0.06 and 1%. Arteriovenous fistula formation may be associated with simultaneous puncture and catheterization of the vein and artery due to adverse anatomy.

NEURAL DAMAGE

This occurred in 0.18% of angioplasties by the femoral approach in our experience, but was not always permanent. The risk and consequences are much greater if neural damage occurs during axillary and brachial artery puncture.

ABSCESS

An abscess may form if an infection occurs at the site of puncture particularly in the presence of a haematoma.

Complications at the site of angioplasty (Table 1.5)

OCCLUSION

Converting a stenosis to an occlusion whilst attempting to cross the lesion is particularly likely if dissection occurs with the wire and catheter passing in an antegrade direction. This is avoided by very gentle manipulation of wires and catheters. If there are large collaterals which are unaffected, the patient's symptoms may not worsen significantly,

Occlusive:	Dissection Thrombosis
Haemorrhagic:	Perforation Rupture

Table 1.5 *Complications at the angioplasty site*

but at worst a patient with intermittent claudication can be converted to one with rest pain and threatened limb loss. Similarly, damage of large collateral vessels by dissection may cause worsening of the patient's symptoms.

Dissection and reocclusion are more likely to occur if attempts are made to recross a segment of artery that has just undergone balloon dilation, so it is important to keep a wire across the lesion at all times until a satisfactory lumen has been achieved.

Occlusion at the site of angioplasty due to dissection may be managed by redilation with prolonged inflation time (e.g. 2–10 min) if the lumen can be successfully recanalized. If reocclusion occurs due to intimal flaps falling across the lumen successful resection of these flaps with the Simpson atherocath has been reported (Maynar *et al.*, 1989). Alternatively, an intra-arterial stent may be placed. Reocclusion within the first month of angioplasty can be treated with thrombolysis with repeat angioplasty of any residual stenosis.

HAEMORRHAGE

Perforation of an occluded segment of artery rarely causes any significant haemorrhage. However, perforation in a stenotic artery may cause haemorrhage and a large haematoma to develop with subsequent compression of the artery which may require surgical intervention.

Rupture of the artery during balloon dilation has been reported in the suprainguinal vessels where haemorrhage may be catastrophic. This is fortunately rare (0.3%) (Zetler, 1978) and may be due to overdilation by using a balloon diameter too large for the vessel or excessive pressure. Severe pain during inflation is a warning sign, but this is not always present. It has been proposed that long-term steroid therapy may predispose to weakening of the arterial wall and consequently rupture (Lois *et al.*, 1985).

Immediate control of bleeding is necessary and may be achieved by:

1 Balloon inflation proximal to the site of rupture while the patient is transferred for surgery
2 Prolonged balloon inflation across the rupture and

then transfer for surgery if bleeding has not stopped (Smith and Cragg, 1989)

3 Occlusion of proximal artery with steel coils prior to femorofemoral by-pass in theatre (Jensen *et al.*, 1985)

Complications distal to the site of percutaneous transluminal angioplasty (Table 1.6)

Spasm is recognized by concentric smooth narrowing of arteries previously widely patent. It occurs more commonly in the infrapopliteal arteries and can be prevented by using prophylactic antispasmodic agents or by infusing antispasmodic agents once it has occurred. When manipulating guidewires in such small vessels it is unwise to allow much movement of the tip of the guidewire as this provokes spasm. In such small arteries, especially in female patients, prevention by administering antispasmodic agents is advisable. A wide range of drugs is available (see Table 1.7) which take effect rapidly when injected directly into the relevant artery.

Distal embolization is one of the most commonly encountered complications. Acute occlusions have a high incidence of embolization and should be treated first with fibrinolysis. Iliac occlusions >6 cm in length have a high incidence of embolization (Cumberland, 1982). Large emboli usually lodge at the bifurcation of the popliteal artery. These can be treated by:

Spasm
Embolization

Table 1.6 *Complications distal to the percutaneous transluminal angioplasty site*

Antispasmodic drug	Dose	Route
Tolazoline	25 mg	intra-arterial
Isosorbide dinitrate	200–500 μg	intra-arterial
Nifedipine	10 mg	oral
	0.1–0.2 mg	intra-arterial
Lignocaine	100 mg	intra-arterial

Table 1.7 *Antispasmodic drugs*

1 Aspiration thromboembolectomy with a large bore nontapered catheter, for example, 8 or 9 Fr coronary guiding catheter or commercially available aspiration catheter, and a 50 ml syringe (Fig. 1.29). To avoid thrombus getting caught in

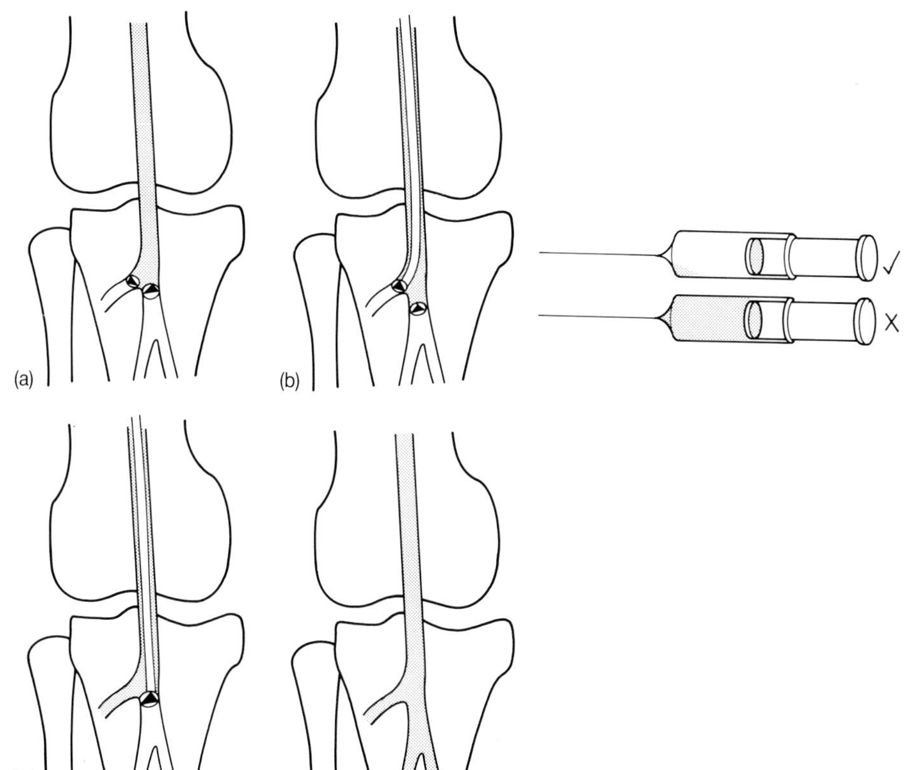

Fig. 1.29 Embolization following percutaneous transluminal angioplasty (PTA)
(a) The emboli have lodged at the tibial bifurcation. (b) An 8 Fr wide bore catheter is passed down to them and the emboli are aspirated. Little blood is withdrawn into the syringe if emboli are lodged at the tip of the catheter. (c) The procedure is repeated, directing the catheter at the remaining emboli seen on a repeat angiogram. (d) Patency can be achieved after aspiration of the emboli.

the check flow valve or sheath, a sheath with a detachable valve is useful. Careful manipulation across the angioplasty site with a floppy tipped guidewire is essential, as a wire cannot be kept in place.

2 Thrombolysis, if the thrombus is thought to be relatively fresh.
3 Surgical embolectomy may be necessary if the fragment is very large and cannot be withdrawn by method 1. Multiple small emboli causing trash foot are very unusual following PTA.

Systemic complications (see Table 1.8)

Cholesterol embolization is due to extensive catheter manipulation in severe arteriopaths. Warning signs during the procedure are development of backache and increasing patient restlessness. It is recognized by livedo reticularis extending from the umbilicus to the toes and initially preservation of peripheral pulses despite marked peripheral ischaemia. Patients may develop renal failure and there is a high mortality rate (Gaines *et al.*, 1988).

Contrast medium related	Mild allergy Anaphylactic shock Renal impairment
Catheter related	Cholesterol embolization Mesenteric infarction Cerebrovascular accident Myocardial infarction Septicaemia Hypotension due to blood loss or excessive use of vasodilator

Table 1.8 *Systemic complications*

RESULTS

The results of aortoiliac PTA are given in Table 1.9, of femoro popliteal PTA in Table 1.10 and of tibial PTA in Table 1.11.

Study	Number of cases	Initial success	Follow-up period (years)	Patency rate
Aortoiliac stenoses				
Kadir *et al.* (1983)	130	97%	1	91%
			3	89%
Gallino *et al.* (1984)	134	95%	5	83%
van Andel *et al.* (1985)	194	96%	7	90%
In der Maur *et al.* (1990)	157	93%	5	84%
Stokes *et al.* (1990)	70 (diabetics)	94%	5	70% (claudicants) 29% (limb salvage)
Iliac occlusions				
Colapinto *et al.* (1986)	64	78%	4	78%

Table 1.9 *Results of aortoiliac percutaneous transluminal angioplasty*

Study	Number of cases	Initial success	Follow-up period (years)	Patency rate	Length of lesion (cm)	
Femoral stenoses						
Krepel *et al.* (1985)	127	92%	5	77%	<2	
				54%	>2	
Murray *et al.* (1987)	116	85%	4.5	67%	<7	
				23%	>7	
In der Maur *et al.* (1990)	55	93%	5	73%		
Stokes *et al.* (1990)	41 (diabetics)	98%	5	60%	(claudicants)	
				8%	(limb salvage)	
Femoral occlusion						
Krepel *et al.* (1985)	37	89%	1	93%	<3	
		26%	1	50%	>3	
Murray *et al.* (1987)	77	74%	4.5	73%		
Zeitler (1990)	58		5	70%	<3	Claudicants
	104		5	65%	3–10	
	59		5	66%	<3	Rest ischaemia
	88		5	54%	3–10	

Table 1.10 *Result of femoropopliteal percutaneous transluminal angioplasty*

Study	Number of cases	Initial success	Follow-up period (years)	Patency rate
Crural PTA				
Tamura *et al.* (1982)	34	85%	2	57%
Schwarten *et al.* (1988)	114	88%	2	86%
Horvath *et al.* (1990)	103	96%	3	65%

Table 1.11 *Results of tibial percutaneous transluminal angioplasty (PTA)*

NEW DEVICES IN ANGIOPLASTY

Readers will be aware of the large number of new devices which have been developed as an adjunct to or a replacement for balloon dilation. It is not appropriate to discuss these in detail in this text, as most of them are still undergoing evaluation and their precise role is yet to be determined.

The aim of these new devices is to try to improve the results of balloon dilation, particularly with respect to long-term patency, but also to improve the immediate technical success of transluminal angioplasty.

Of the mechanical devices, some are used exclusively to recanalize total occlusions, for example, the Rotational Transluminal Angioplasty Catheter System (ROTACS) which is a slowly rotating, battery driven catheter, which takes a guidewire which is left across the occlusion once it has been successfully recanalized and standard balloon dilation is subsequently performed. Some mechanical devices not only recanalize occlusions but remove some of the occluding material by reducing it to microscopic particles, for example the Kensey catheter (Wholey and Jarmolowski, 1989). Other mechanical devices remove the obstructing material but pass over a guidewire, which means that they are most useful in stenoses and occlusions when a guidewire has already successfully recanalized the occlusion, for example the Transluminal Extraction Catheter (TEC), which is suitable for any length of disease and the Simpson atherocath which is most useful in short eccentric and calcified stenoses. Intravascular stents are mechanical devices which maintain a vascular lumen by acting as a scaffold, and their use is indicated when technical success is compromised either due to extrinsic compression of the vessel (e.g. superior vena cava (SVC) syndrome) or intimal flaps which threaten to occlude the lumen following balloon dilation (Becker *et al.*, 1990). Stents have been used when balloon dilation is inadequate, for example when there is a residual pressure gradient across an iliac stenosis, in recurrent stenoses following previous balloon dilation and in venous outflow stenoses in dialysis shunts which are very fibrous and can resist balloon dilation (Günther *et al.*, 1989). There may also be a role for stenting after recanalization of iliac occlusions in an effort to improve patency rates (Vorwerk and Guenther, 1990), but long-term follow-up to confirm this is awaited. In iliac arteries the risk of thrombosis following stent insertion is low but in the femoral arteries the results have been disappointing, re-occlusion of the stented segment being relatively common (Zollikofer *et al.*, 1991).

Laser systems are also used to ablate atheroma and create a channel, but again these are experimental devices, and the reader is referred to more detailed articles on these subjects (Moore and Ahn, 1989; Belli and Cumberland, 1990). Although it has been shown that lasers can recanalize 50% of occlusions in which conventional guidewire methods have failed, the increase in numbers is small (Cumberland and Belli, 1989; Levy *et al.*, 1989). Several randomized trials comparing laser-assisted angioplasty and conventional angioplasty methods in the treatment of peripheral vascular disease are currently underway, and so far report no differences either in the rate of recanalization or long-term patency (Belli *et al.*, 1991b).

New laser systems and mechanical devices allow PTA to remain an option in some cases in which conventional methods of PTA have failed, but no improvement in long-term patency has yet been shown. These devices are still being investigated and their future role is yet to be proven.

References

Auster M, Kadir S, Mitchell SE, Williams GM, Perler B, Chang R, White RI Jr (1984) Iliac artery occlusions: Management with intrathrombus streptokinase infusion and angioplasty. *Radiology* **153**: 385–8.

Baudouin CJ, Belli AM, Peck RJ, Cumberland DC (1990) The complication rate of high brachial puncture. *Clinical Radiology* **42**: 277–80.

Becker GJ, Palmaz JC, Rees CR, Ehrman KO, Lalka SG, Dalsing MC, Cikrit DF, McLean GK, Burke DR, Richter GM, Noeldge G, Giarcia O, Waller BF, Castaneda-Zuniga WR (1990) Angioplasty induced dissections in human iliac arteries. Management with Palmaz balloon-expandable intraluminal stents. *Radiology* **176**: 31–8.

Belli AM, Cumberland DC (1990) Lasers in radiology. *Clinical Radiology* **41**: 75–6.

Belli AM, Cumberland DC, Knox A, Procter AE, Welsh CL (1990a) Complication rate of percutaneous peripheral balloon angioplasty. *Clinical Radiology* **41**: 380–3.

Belli AM, Cumberland DC, Procter AE, Welsh CL (1991a) Total peripheral artery occlusion: conventional versus laser thermal recanalization with a hybrid probe in percutaneous angioplasty – results of a randomized trial. *Radiology* **181**: 57–60.

Belli AM, Cumberland DC, Procter AE, Welsh CL (1991b) Follow-up of conventional versus laser thermal percutaneous angioplasty of total femoropopliteal artery occlusions: results of a randomized trial. *Journal of Vascular and Interventional Radiology* **2**: 485–8.

Belli AM, Hemingway AP, Cumberland DC, Welsh CL (1989) Percutaneous transluminal angioplasty of the distal abdominal aorta. *European Journal of Vascular Surgery* **3**: 449–53.

Belli AM, Procter AE, Cumberland DC (1990b) Peripheral vascular occlusions: mechanical recanalization with a metal laser probe after guidewire dissection. *Radiology* **176**: 539–41.

Brewer ML, Kinnison ML, Perler BA, White RI Jr (1988) Blue toe syndrome: treatment with anticoagulants and delayed percutaneous transluminal angioplasty. *Radiology* **166**: 31–6.

Carver RA, Cumberland DC (1987) Use of cobra-shaped catheter to effect recanalisation of femoro-popliteal occlusions. *Journal of Interventional Radiology* **2**: 97–8.

Colapinto RF, Stronell RD, Johnston WK (1986) Transluminal angioplasty of complete iliac obstructions. *Radiology* **146**: 859–62.

Cumberland DC (1982) Percutaneous angioplasty in complete iliac occlusions. *VASA* **II**: 297–300.

Cumberland DC, Belli AM (1989) Laser assisted angioplasty. In: *Pros and Cons in PTA and Auxilliary Methods*, pp. 91–7. Edited by Zeitler E, Seyferth W. Springer-Verlag, Berlin, Heidelberg.

Fergusson DJ, Kamada RO (1981) Percutaneous entry of the brachial artery for left heart catherization using a sheath. *Catheterization and Cardiovascular Diagnosis* **7**: 111–14.

Fontaine R, Kim M, Kieny R (1954) Die Chirurgische Behandlung der periferen Durchblutungsstörungen. *Helvetica Chirurgica Acta* **21**: 499–533.

Gaines PA, Cumberland DC, Kennedy A, Welsh CL, Moorhead P, Rutley MS (1988) Cholesterol embolization: A lethal complication of vascular catherization. *Lancet* **i**: 168–70.

Gaines PA, Reidy JF (1986) Percutaneous high brachial aortography: a safe alternative to the translumbar approach. *Clinical Radiology* **37**: 595–7.

Gallino A, Mahler F, Probst P, Nachbur B

(1984) Percutaneous transluminal angioplasty of the arteries of the lower limbs: A 5 year follow-up. *Circulation* **70:** 619–23.

GARDINER GA, MEYEROVITZ MF, STOKES KR, CLOUSE ME, HARRINGTON DP, BETTMANN MA (1986) Complications of transluminal angioplasty. *Radiology* **159:** 201–8.

GLANZ S, GORDON D, BUTT KMH, HONG J, LIPKOWITZ GS (1987) The role of percutaneous angioplasty in the management of chronic haemodialysis fistulas. *Annals of Surgery* **206:** 777–81.

GRIER D, HARTNELL G (1990) Percutaneous femoral artery puncture: practice and anatomy. *British Journal of Radiology* **63:** 602–4.

GÜNTHER RW, VORWERK D, BOHNDORF D, KLOSE KC, KISTLER D, MANN H, SIEBERTH HG, EL-DIN A (1989) Venous stenoses in dialysis shunts: Treatment with self-expandable metallic stents. *Radiology* **170:** 401–5.

HESSEL SJ, ADAMS DF, ABRAMS HL (1981) Complications of angiography. *Radiology* **138:** 273–381.

HORVATH W, OERTL M, HAIDINGER D (1990) Percutaneous transluminal angioplasty of crural arteries. *Radiology* **177:** 565–69.

IN DER MAUR GAP, DE BOO T, BOEVÉ J, KERDEL MC, BRAAKENBURG BA (1990) Angioplasty of the iliac and femoral arteries. Initial and long-term results in short stenotic lesions. *European Journal of Radiology* **11:** 163–67.

JENSEN SR, VOEGELI DR, CRUMMY AB, TUNIPSEED WD, ACHER CW, GOODSON S (1985) Iliac artery rupture during transluminal angioplasty; Treatment by embolization and surgical bypass. *American Journal of Roentgenology* **145:** 381–2.

KADIR S, WHITE RI JR, KAUFMAN SL, BARTH KH, WILLIAMS GM, BURDICK JF, O'MARA CS, SMITH GW, STONESIFER GL JR, ERNST CB, MINKEN SL (1983) Long-term results of aortoiliac angioplasty. *Surgery* **94:** 10–14.

KREPEL VM, VAN ANDEL GJ, VAN ERP WFM, BRESLAU PJ (1985) Percutaneous transluminal angioplasty of the femoropopliteal artery: Initial and long-term results. *Radiology* **156:** 325–28.

KUMPE DA, ZWERDLINGER S, GRIFFIN DJ (1988) Blue digit syndrome: treatment with percutaneous transluminal angioplasty. *Radiology* **166:** 37–44.

LEVY JM, HESSEL SJ, HORSLEY WW, COOK GC, DICKEY JE (1989) Value of laser assisted angioplasty in the community hospital. *Radiology* **170:** 1017–18.

LOIS JF, TAKIFF H, SCHECHTER MS, GOMES AS, MACHLEDER HI (1985) Vessel rupture by balloon catheters complicating chronic steroid therapy. *American Journal of Roentgenology* **144:** 1073–4.

MAYNAR M, REYES R, CABRERA V, ROMAN M, PULIDO JM, CASTANEDA F, LETOWNEAU JG, CASTANEDA-ZUNIGA WR (1989) Percutaneous atherectomy as an alternative treatment for postangioplasty obstructive intimal flaps. *Radiology* **170:** 1029–31.

MOORE WS, AHN SS (1989) *Endovascular Surgery*. W.B. Saunders Company, Philadelphia.

MURRAY RR, HEWES RC, WHITE RI, MITCHELL SE, AUSTER M, CHANG R, KADIR S, KINNISON M, KAUFMAN SL (1987) Long segment femoropopliteal stenoses: is angioplasty a boon or a bust? *Radiology* **162:** 473–76.

RING EJ, FREIMAN DB, MCLEAN GK, SCHWARZ W (1982) Percutaneous recanalization of common iliac artery occlusions: An unacceptable complication rate? *American Journal of Roentgenology* **139:** 587–9.

SCHWARTEN DE, CUTCLIFF WB (1988) Arterial occlusive disease below the knee: Treatment with percutaneous transluminal angioplasty performed with low-profile catheters and steerable guide wires. *Radiology* **167:** 71–74.

SMITH TP, CRAGG AH (1989) Non surgical treatment of iliac artery rupture following angioplasty. *Journal of Interventional Radiology* **4:** 16–18.

SONES FM JR, SHIRLEY EK (1962). Cine coronary arteriography. *Modern Concepts of Cardiovascular Disease* **31:** 735–8.

SPIJKERBOER AM, SCHOLTEN FG, MALI WP, VAN SCHAIK JPJ (1990) Antegrade puncture of the femoral artery: Morphologic study. *Radiology* **176:** 57–60.

STOKES KR, STRUNK HM, CAMPBELL DR, GIBBONS GW, WHEELER HG, CLOUSE ME (1990) Five year results of iliac and femoropopliteal angioplasty in diabetic patients. *Radiology* **174:** 987–92.

TAMURA S, SNIDERMAN KW, BEINART C, SOS TA (1982) Percutaneous transluminal angioplasty of the popliteal artery and its branches. *Radiology* **143:** 645–48.

TØNNESEN KH, SAGER P, KARLE A, HENRIKSEN L, JØRGENSEN B (1988) Percutaneous transluminal angioplasty of the superficial femoral artery by retrograde catheterization via the popliteal artery. *Cardiovascular and Interventional Radiology* **11:** 127–31.

VAN ANDEL GJ, VAN ERP WFM, KREPEL VM, BRESLAU PJ (1985) Percutaneous transluminal dilatation of the iliac artery: Long-term results. *Radiology* **156:** 321–23.

VITEK JJ (1989) Subclavian artery angioplasty and the origin of the vertebral artery. *Radiology* **170:** 407–9.

VORWERK D. GUENTHER RW (1990) Mechancial revascularization of occluded iliac arteries with use of self-expandable endoprostheses. *Radiology* **175:** 411–15.

WHOLEY MH, JARMOLOWSKI CR (1989) New reperfusion devices: The Kensey catheter, the atherolytic reperfusion wire device, and the transluminal extraction catheter. *Radiology* **172:** 947–52.

ZEITLER E (1978) Complications in and after PTA. In: *Percutaneous Vascular Recanalisation*, pp. 120–5. Edited by Zeitler E, Grüntzig AR, Schoop W. Springer-Verlag, New York.

ZEITLER E (1990) Percutaneous transluminal angioplasty of the femorotibial arteries. In: *Interventional Radiology* pp. 617–24. Edited by Dondelinger RF, Rossi P, Kurdziel JC and Wallace S. Thieme Medical Publishers, New York.

ZOLLIKOFFER CL, ANTONUCCI F, PFYFFER M, REDHA F, SALOMONOWITZ E, STUCKMANN G, LARGIADÈR I, MARTY A (1991) Arterial stent placement with use of the Wallstent: Midterm results of clinical experience. *Radiology* **179:** 449–56.

CHAPTER 2

Thrombolysis

Roger H. S. Gregson

Introduction 30

Indications for thrombolysis 30

Indications for vascular surgery 37

Indications for conservative management 40

Contraindications to thrombolysis 40

Prelysis clinical and radiological patient management 40

Prelysis problems 44

Thrombolysis management 47

Thrombolysis problems 48

Postlysis clinical and radiological patient management 53

Postlysis problems 54

Haematological management 55

Fibrinolytic drugs 55

Complications of thrombolysis 57

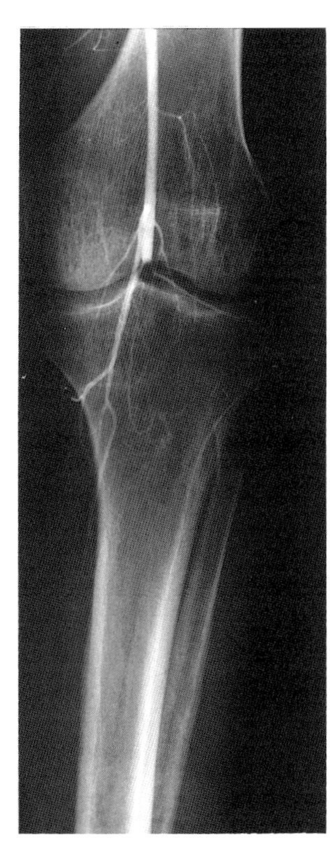

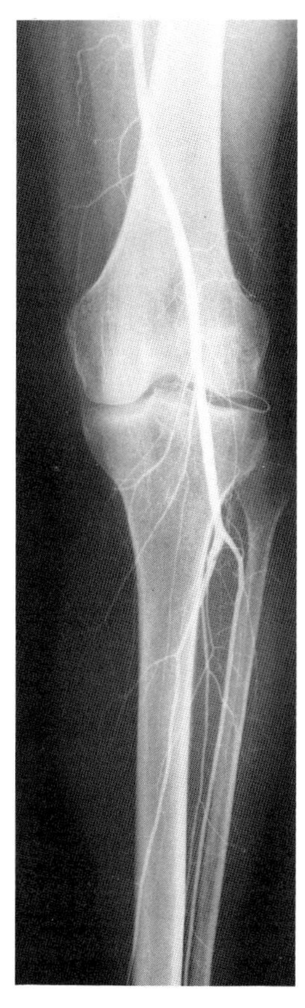

Results of thrombolysis 58

Recent advances in thrombolysis 58

References 59

Introduction

Fibrinolytic therapy has been used intermittently in the treatment of a number of different thrombotic diseases for over 30 years, including acute myocardial infarction, acute upper and lower limb arterial ischaemia, acute renal and mesenteric artery occlusion, deep venous thrombosis in the arm and leg, and pulmonary embolism, but it is in vogue at present for the treatment of both acute myocardial infarction and acute lower limb arterial ischaemia in particular.

Some of the first reported cases in the 1960s (Cotton *et al.*, 1962; McNicol *et al.*, 1963) described the successful use of intra-arterial streptokinase infusions in the treatment of acute popliteal artery occlusion. However, the series published over the next 10 years all reported the use of intravenous infusions of high-dose streptokinase following a loading dose which was required to neutralize anti-streptococcal antibodies (Amery *et al.*, 1970; LeVeen and Diaz, 1972; Fiessinger *et al.*, 1978). The loading dose of streptokinase varied between 200000 and 1000000 units, but the intravenous infusion was usually 100000 units/hour.

Successful lysis of thrombus occurred in 30–45% of patients. Against this low rate of success had to be balanced a significant number of complications, which included both haemorrhagic complications in 10–20% of patients as well as thromboembolic complications in about 10% of patients. A mortality rate of more than 30% was also reported by Amery *et al.* (1970) with this treatment.

In order to minimize the haemorrhagic complications produced by the systemic fibrinolytic effect of an intravenous infusion of high-dose streptokinase, Dotter *et al.* (1974) developed a technique whereby an intra-arterial infusion of a low dose of streptokinase could be delivered by a catheter positioned with its tip into or near the acute occlusion. Following this description in a small series of 17 patients, the technique of low-dose intra-arterial thrombolytic therapy has become widely used, initially in the USA and Europe, but more recently in the UK as well.

Indications for thrombolysis

Thrombolysis appears to be the most appropriate form of treatment for the majority of patients with critical lower limb ischaemia, because they are likely to have a poor clinical outcome with conventional surgical treatment (Blaisdell *et al.*, 1978; McPhail *et al.*, 1983; Dale, 1984). However, not all patients presenting with acute, subacute or acute on chronic arterial ischaemia of the lower limb are suitable for treatment with intra-arterial thrombolytic therapy. For some a surgical procedure is undoubtedly the treatment of choice, for others conservative management is best and in a few thrombolytic therapy is contraindicated.

The indications for intra-arterial thrombolytic therapy in patients with critical lower limb ischaemia (which is defined as rest pain in an ischaemic leg with a Doppler pressure of less than 40 mmHg or a Doppler pressure of less than 60 mm Hg, when there is associated tissue necrosis) depends upon both the clinical and radiological diagnosis and therefore includes patients with:

1. Distal arterial thrombosis extending from the superficial femoral or popliteal arteries into the tibial arteries, including thrombosis of a popliteal artery aneurysm (Fig. 2.1).
2. Distal arterial embolus involving the popliteal and tibial arteries (Fig. 2.2).
3. Proximal arterial thrombosis extending from the iliac arteries into the femoral arteries with no distal run off in the popliteal or tibial arteries (Fig 2.3).
4. Thrombosis of a synthetic or vein graft (Figs 2.4 and 2.5),

but also patients with:

5. Proximal or distal arterial thrombosis with patent distal run off vessels, when there is likely to be a high risk from reconstructive vascular surgery (Figs 2.6 and 2.7).
6. Proximal arterial embolus involving the iliac or femoral arteries, if there is likely to be a low success rate from thromboembolectomy surgery (Figs 2.8 and 2.9).
7. (Proximal or) distal arterial thrombosis with patent distal run off vessels, when there is a reason to withhold vascular surgery (Figs 2.6 and 2.7).
8. Thrombosis at the site of a recent angioplasty (Fig. 2.10).

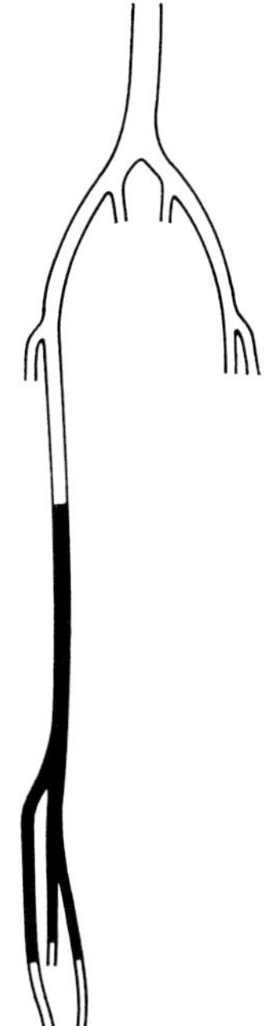

Fig. 2.1 Diagram of distal superficial femoral arterial thrombosis extending into popliteal and tibial arteries

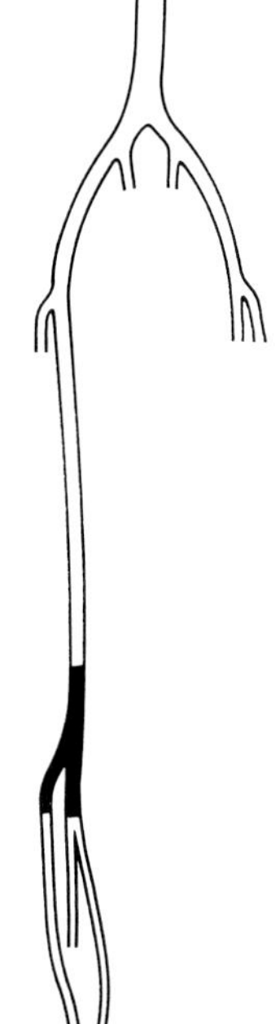

Fig. 2.2 Diagram of distal popliteal arterial embolus involving tibial arteries

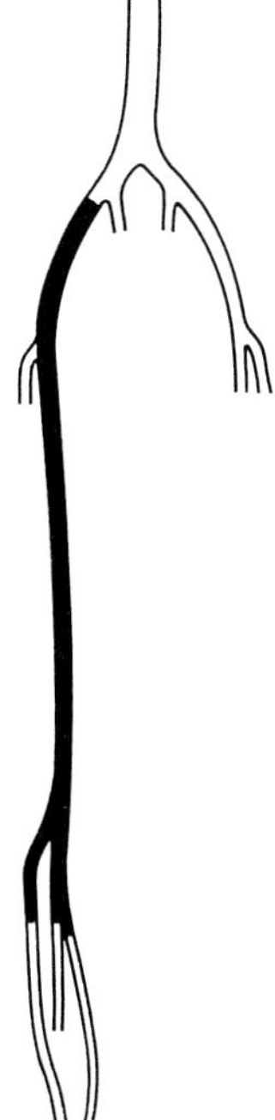

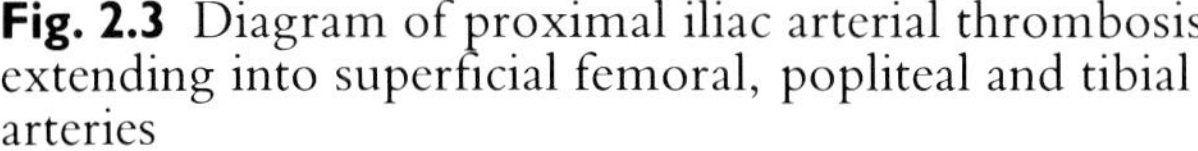

Fig. 2.3 Diagram of proximal iliac arterial thrombosis extending into superficial femoral, popliteal and tibial arteries

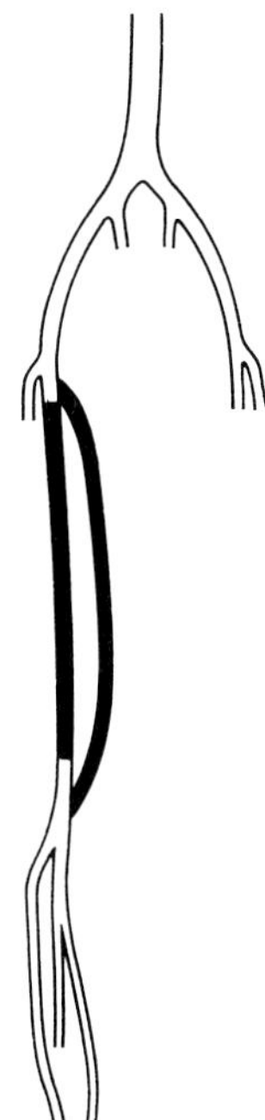

Fig. 2.4 Diagram of femoro-popliteal graft thrombosis

Fig. 2.5 Diagram of femoro-femoral graft thrombosis

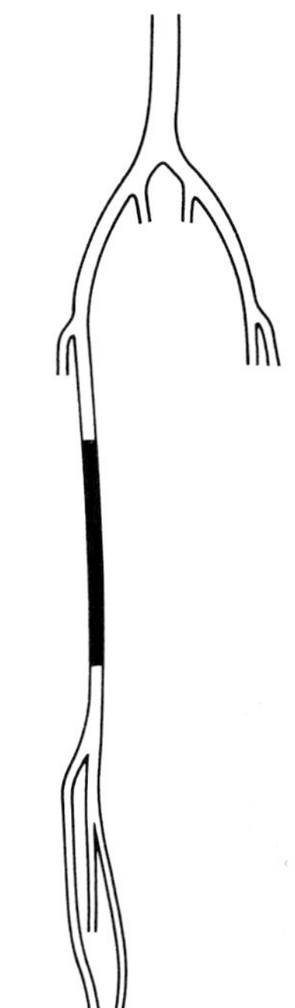

Fig. 2.6 Diagram of superficial femoral arterial thrombosis

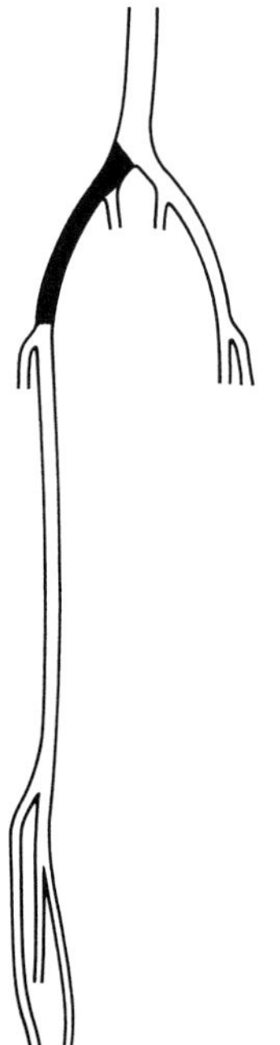

Fig. 2.7 Diagram of iliac arterial thrombosis

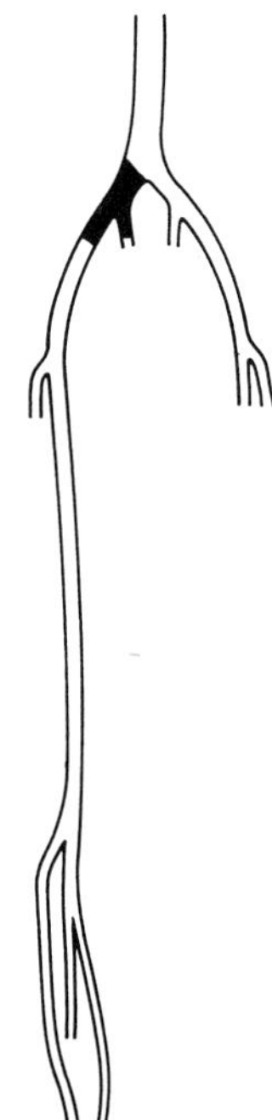

Fig. 2.8 Diagram of common iliac arterial embolus

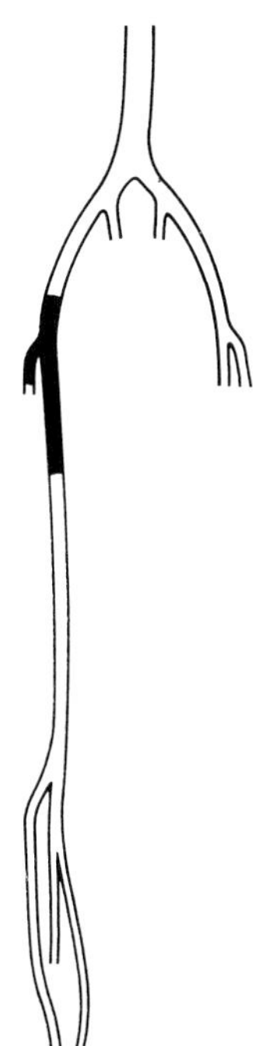

Fig. 2.9 Diagram of common femoral arterial embolus

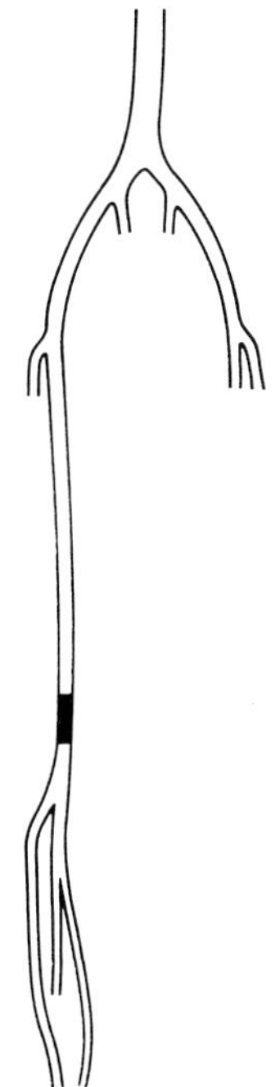

Fig. 2.10 Diagram of popliteal artery thrombosis after angioplasty

Intra-arterial thrombolysis is the treatment of choice for patients presenting within 30 days with an acute proximal or distal arterial thrombosis, who have severe ischaemia with normal muscle power and intact sensation on clinical examination, but occlusion of the popliteal and tibial arteries demonstrated by arteriography (Fig. 2.11). This is because conventional reconstructive surgery is not possible and mortality rates of 15–50% and amputation rates of 20–60% have been reported with thromboembolectomy (Blaisdell *et al.*, 1978), particularly if there is significant atherosclerosis in the lower limb vessels.

Intra-arterial thrombolysis is the treatment of choice for patients presenting within 48–72 hours with acute distal arterial emboli, because it is difficult to remove all the thrombus from the tibial arteries with a Fogarty catheter, which can easily damage these small vessels. Thrombolysis is also the treatment of choice for patients presenting after 2–5 days with acute proximal or distal arterial emboli, because both underlying atherosclerosis in the arterial wall and embolic thrombus damage to the intima predispose to recurrent thrombosis after a successful thromboembolectomy (Fig. 2.12).

Intra-arterial thrombolysis is the treatment of choice for patients presenting with thrombosis of a vein or synthetic graft, because graft thrombectomy is unlikely to be successful as it does not treat the underlying cause of the graft occlusion (Fig. 2.13). Early graft failure is usually due to the development of intimal hyperplasia stenoses and late graft failure to the progression of atherosclerotic disease.

Thrombolysis is best in patients with an arterial thrombosis who have a high risk with surgical treatment due to a recent myocardial infarction. It should also be considered in the young patient with arterial thrombosis under the age of 40 years, so that the native vessels can be preserved, leaving by-pass grafting for the future, if necessary. Thrombolysis is also the treatment of choice for patients who develop thrombosis at the site of angioplasty (or arterial puncture), particularly if their leg deteriorates clinically (Fig. 2.14).

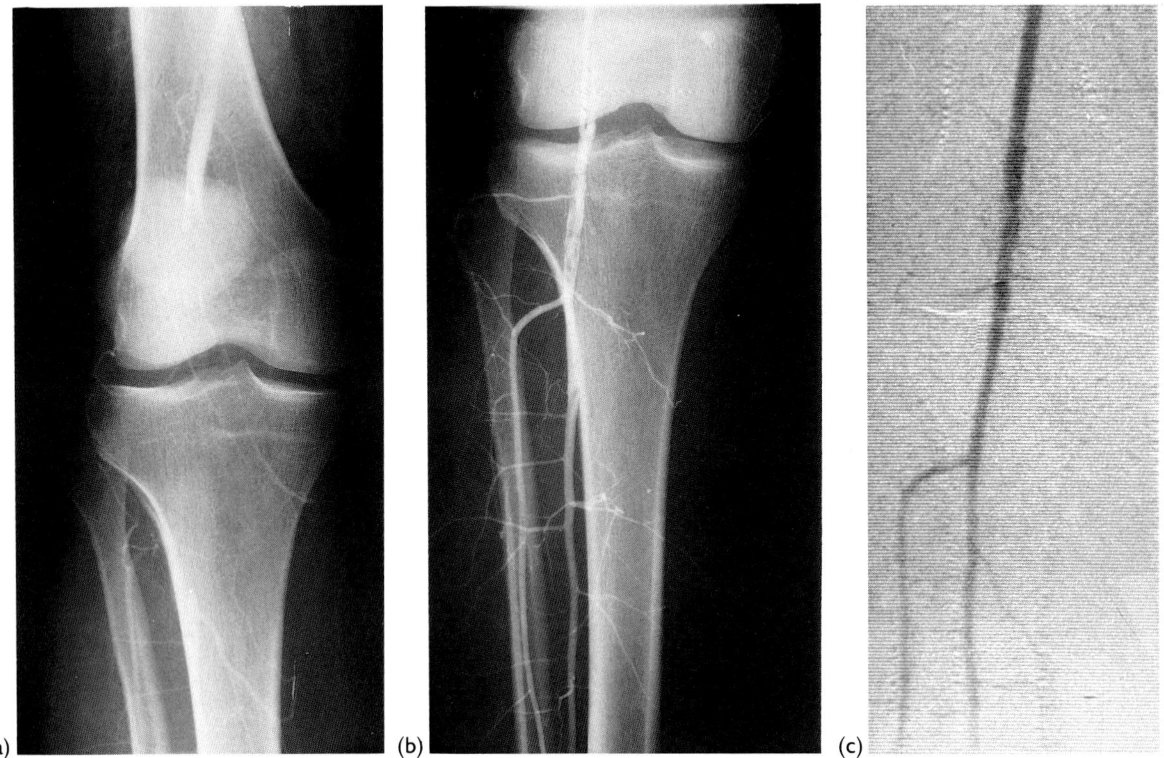

Fig. 2.11
X-rays showing (a) distal right popliteal arterial thrombosis extending into tibial arteries; (b) during lysis with 5 Fr catheter in the thrombus; (c) one month after successful lysis.

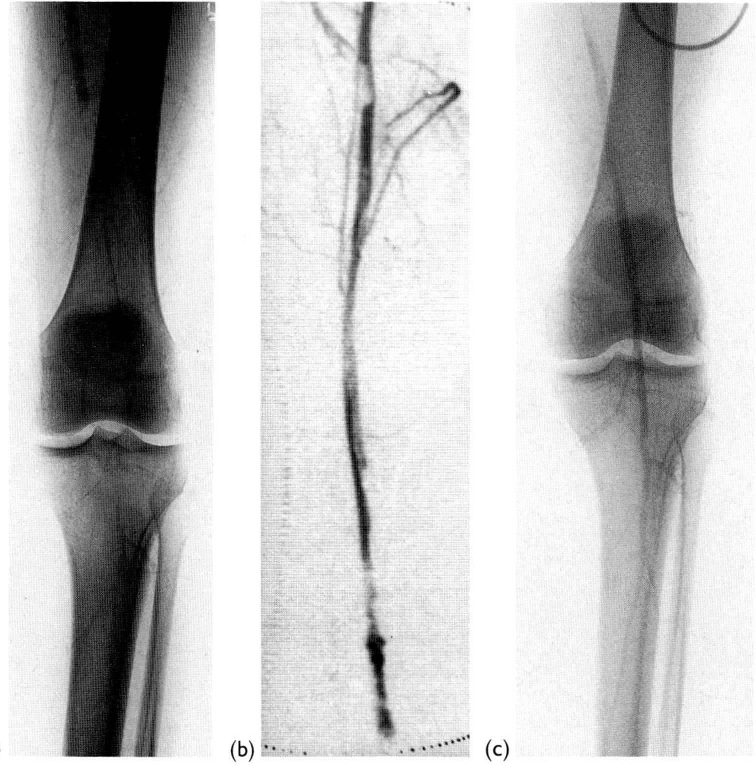

Fig. 2.12
(a–c) X-rays showing (a) distal left superficial femoral arterial embolus extending into popliteal and tibial arteries; (b) before lysis with 4 Fr catheter in the thrombus; (c) immediately after successful lysis.

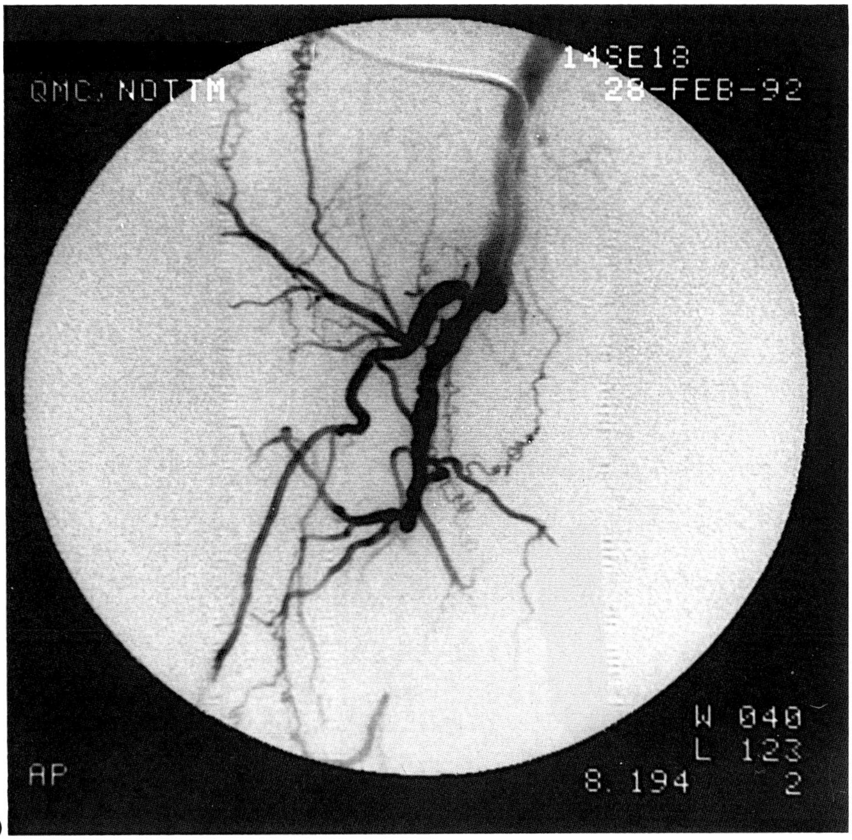

(a)

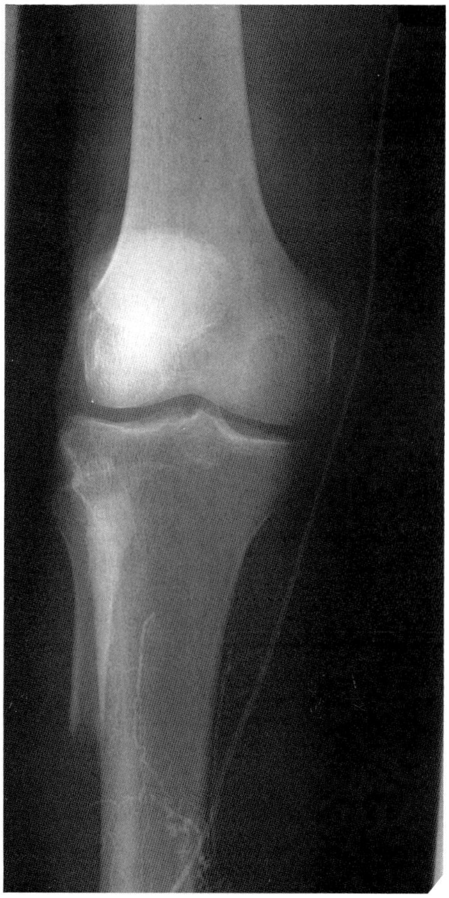

(b)

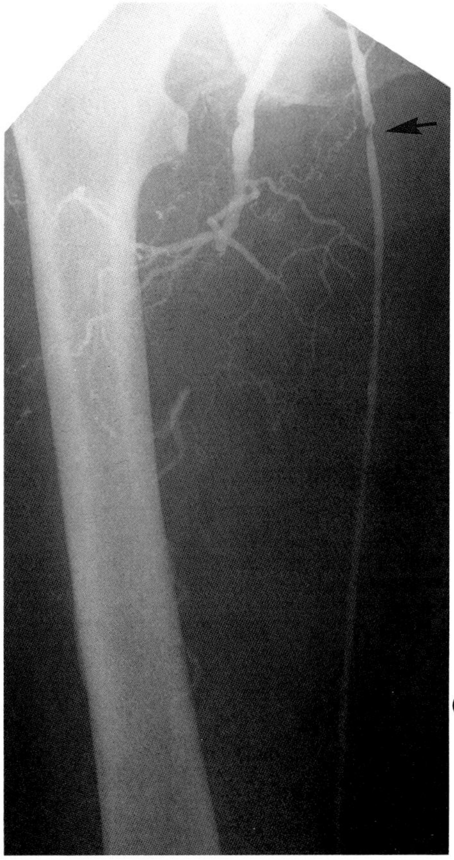

(c)

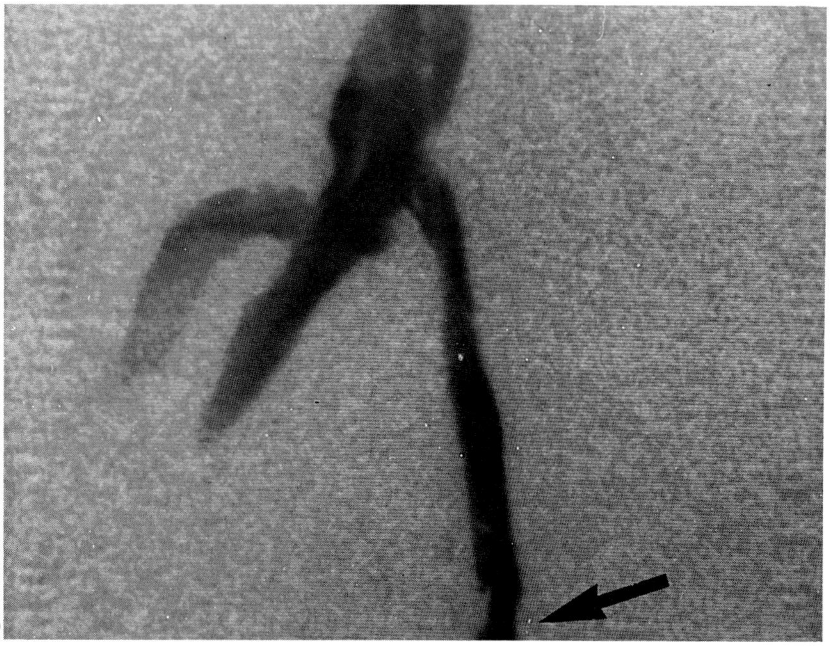

(d)

Fig. 2.13
(a–d) X-rays showing (a) right femorodistal vein graft thrombosis (to the common peroneal artery); (b) after successful lysis distally; (c) after successful lysis proximally a vein graft stenosis is revealed; (d) after successful angioplasty.

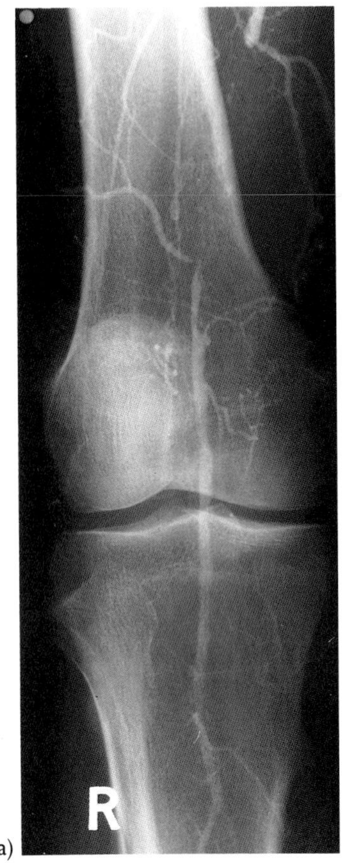

(a)

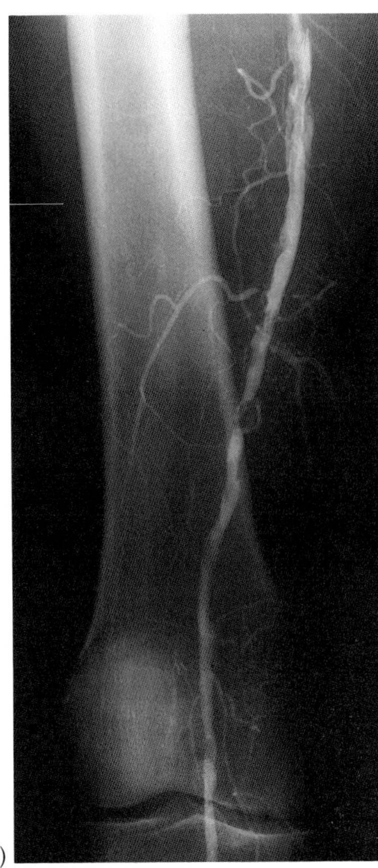

(b)

Fig. 2.14
(a–e) X-rays showing (a) short distal right superficial femoral/popliteal artery occlusion; (b) after recanalization angioplasty; (c) acute rethrombosis at the site of angioplasty 2 weeks later; (d) immediately after successful lysis; (e) 1 month after successful lysis.

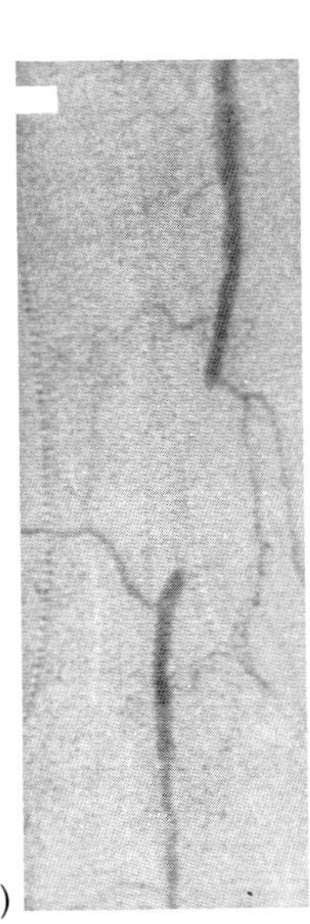

(c)

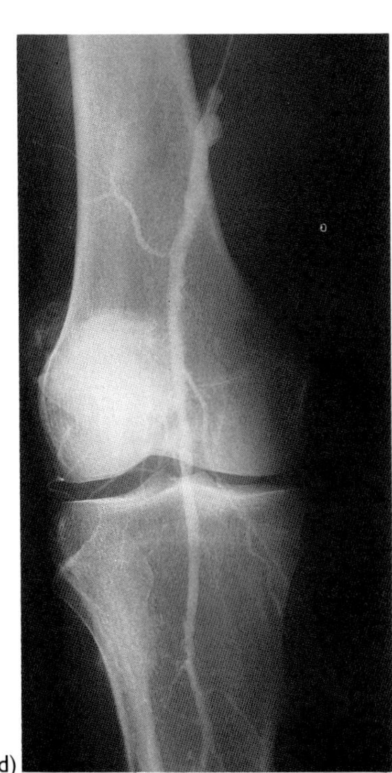

(d)

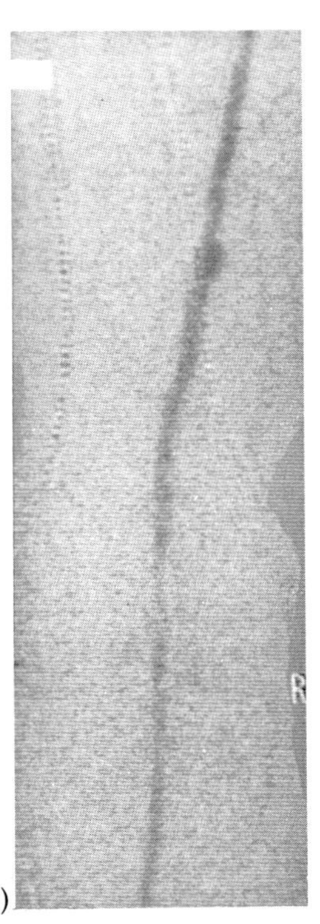

(e)

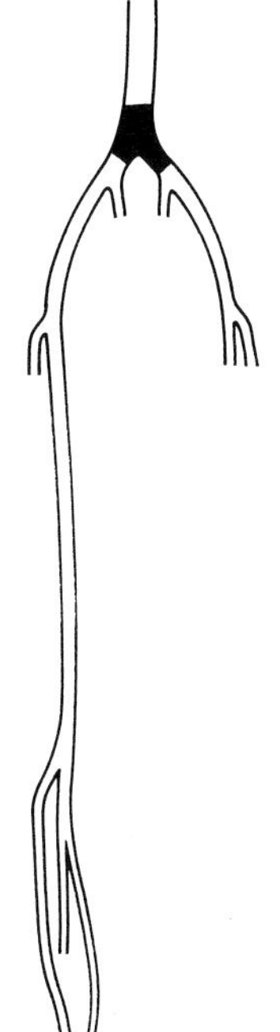

Fig. 2.15 Diagram of aortic bifurcation saddle embolus

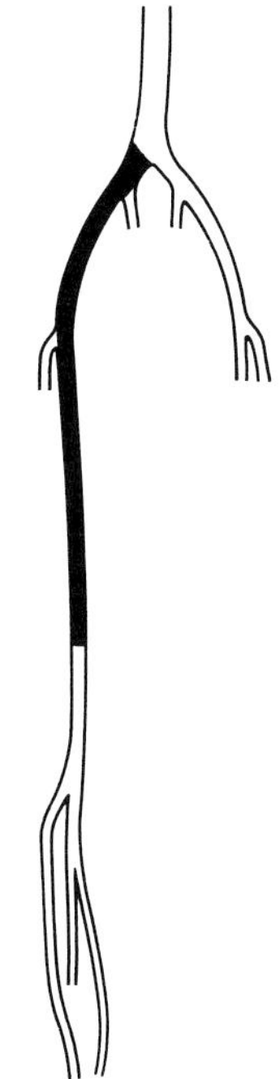

Fig. 2.16 Diagram of proximal iliac arterial thrombosis extending into superficial femoral artery

INDICATIONS FOR VASCULAR SURGERY

The indications for surgery in patients with critical lower limb ischaemia also depend upon both the clinical and radiological diagnosis and therefore includes patients with:

1. Proximal arterial embolus involving the aortic bifurcation, iliac or femoral arteries (Figs 2.8, 2.9 and 2.15).
2. Proximal or distal arterial thrombosis with patent distal run off vessels, if there is insufficient time for intra-arterial thrombolysis (Fig. 2.16).
3. Postoperative thrombosis of a synthetic or vein graft.
4. Failed thrombolysis.

Surgical thromboembolectomy using a Fogarty catheter is the treatment of choice for patients presenting within 48–72 hours with acute proximal arterial emboli because there is a mortality rate of only 12%, and a limb salvage rate of 95% can be achieved in surviving patients (Tawes *et al.*, 1985), particularly if there is no significant atherosclerosis in the lower limb vessels (Fig. 2.17).

Reconstructive vascular surgery with either a synthetic or vein by-pass graft is the treatment of choice for patients presenting with an acute proximal or distal arterial thrombosis, who have severe ischaemia with muscle paralysis and absent sensation on clinical examination, but patent run off vessels demonstrated by arteriography below the occlusion. This is because urgent revascularization of the leg is required within 6–12 hours to prevent tissue necrosis and the length of the occlusion does not allow intra-arterial thrombolytic therapy sufficient time to achieve limb salvage. Thrombolysis in this clinical situation, where there is successful reperfusion of necrotic muscle, can in fact lead to myoglobinuria and the development of acute renal failure.

Further surgery is also best in patients who present with a thrombosed graft within 2–4 weeks of their operation, because thrombolysis is likely to produce bleeding from either the anastomotic suture line of a vein or synthetic graft or haemorrhage through the interstices of a synthetic graft under these circumstances. Amputation may become necessary in patients where intra-arterial thromboly-

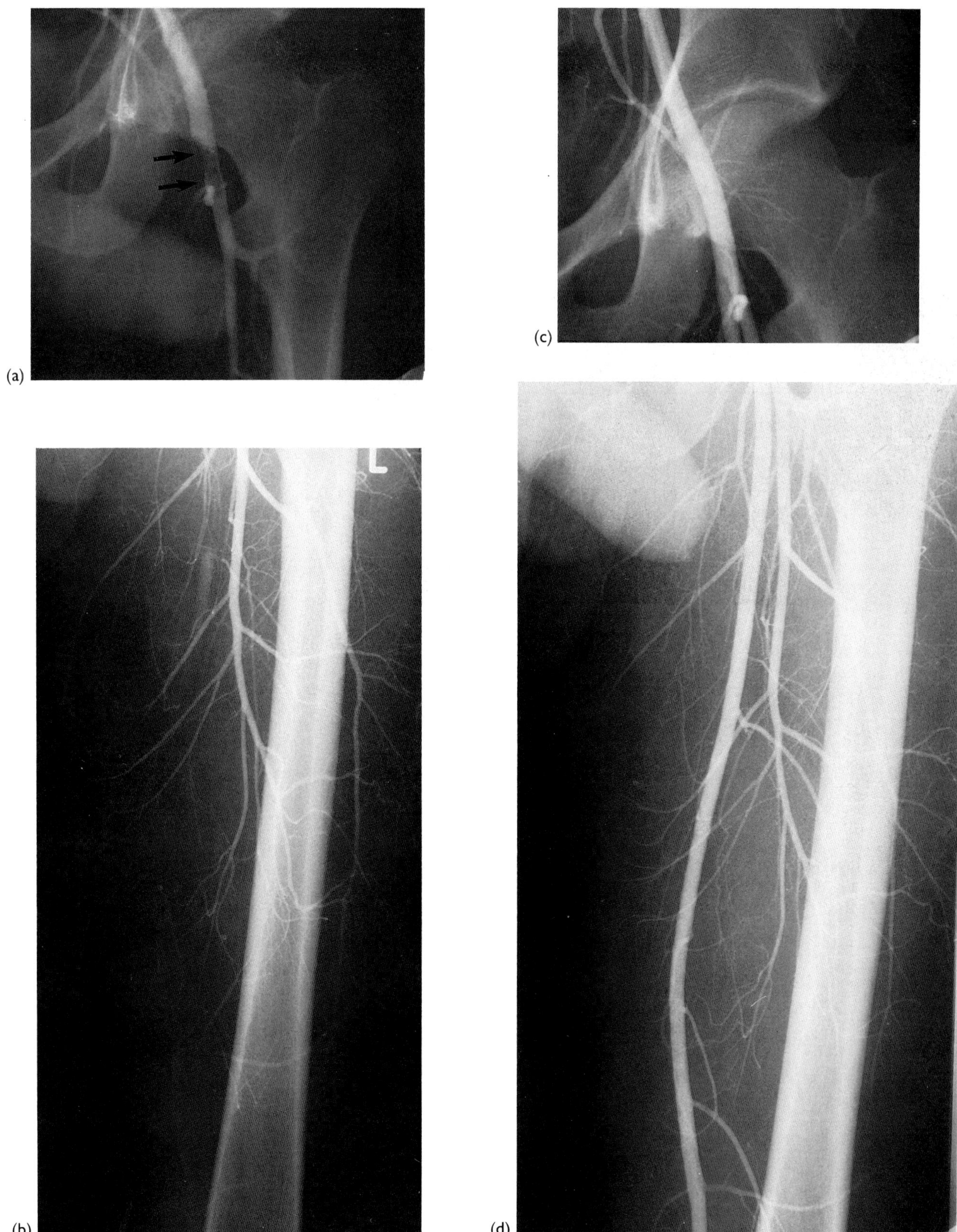

Fig. 2.17
X-rays showing (a,b)proximal left common femoral arterial embolus extending into superficial femoral artery; (c,d) six months after embolectomy.

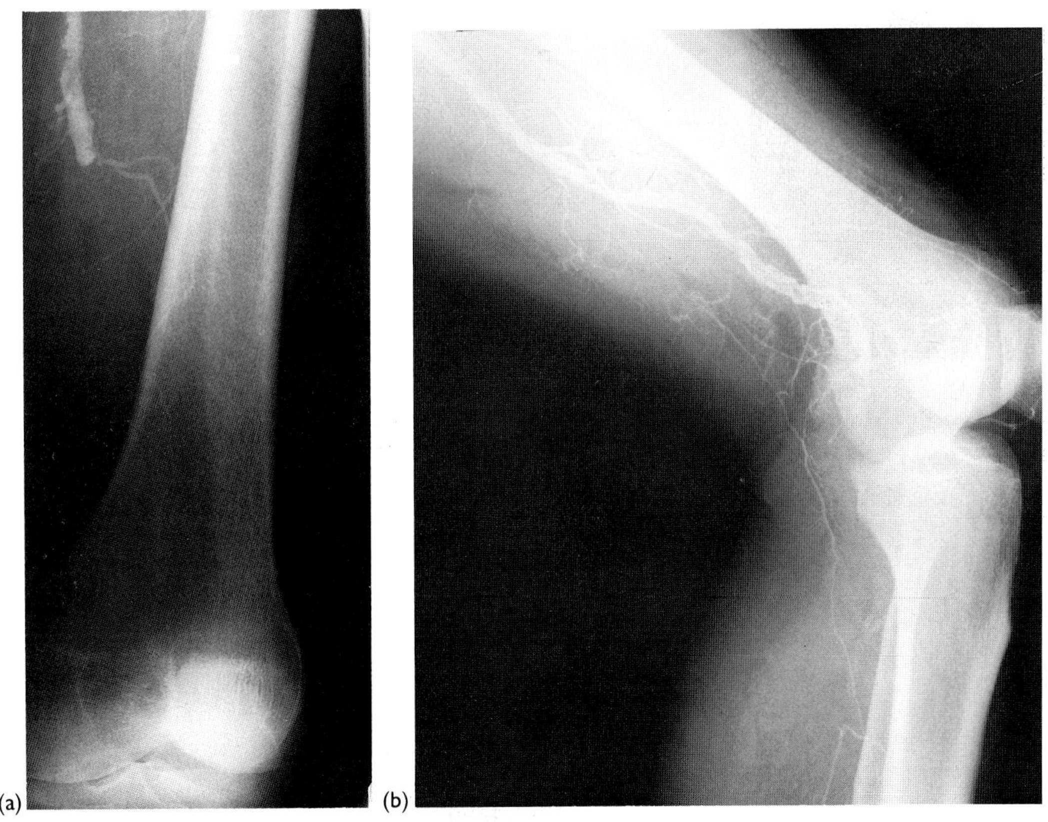

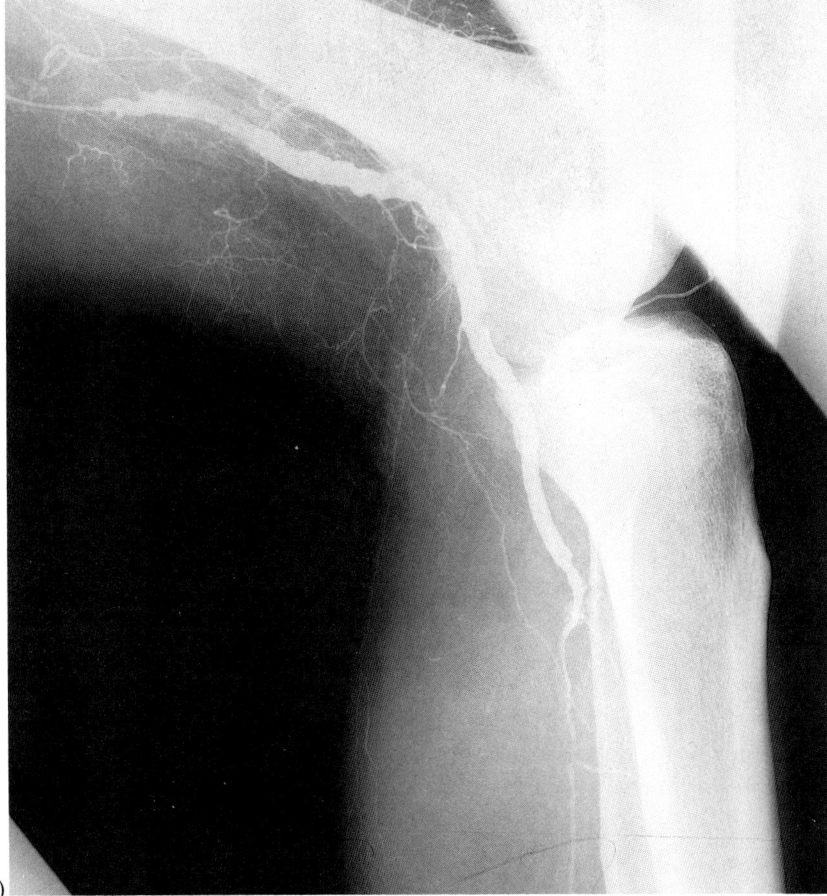

Fig. 2.18
(a–c) X-rays showing (a) distal left superficial femoral arterial thrombosis extending into popliteal and tibial arteries; (b) during lysis with 4 Fr catheter in the thrombus; (c) immediately after successful lysis but amputation was still necessary.

tic therapy and/or reconstructive surgery fails and is inevitable in patients who present with either extensive muscle necrosis or gangrene of the lower limb (Fig. 2.18).

Indications for conservative management

The indications for conservative treatment of patients with critical lower limb ischaemia include patients with either a terminal illness or dementia, because neither thrombolytic nor surgical treatment is suitable. Patients with an associated neoplastic disease, however, should be treated actively with either thrombolysis or surgery.

Contraindications to thrombolysis

The most important potential risk associated with the use of a thrombolytic drug is bleeding and this is present whether it is used systemically in high doses or locally in low doses, as there is still a systemic effect.

The contraindications to intra-arterial thrombolytic therapy (Report of a National Institutes of Health Consensus Development Conference, 1980) in critical lower limb ischaemia include patients with:

1 Active or recent bleeding from any source
2 A stroke within the previous 2–3 months due to the risk of inducing a cerebral haemorrhage at the site of the recent infarct
3 Major surgery within 10–14 days due to the risk of haemorrhage
4 Major trauma within 10–14 days due to the risk of haemorrhage (Fig. 2.19).
5 A known clotting disorder or bleeding diathesis
6 A potential bleeding site, such as a peptic ulcer due to the risk of gastrointestinal haemorrhage or diabetic retinopathy due to the risk of a vitreous haemorrhage
7 Severe hypertension due to the risk of haematoma at the site of arterial puncture
8 Pregnancy and recent delivery
9 Long-term anticoagulant treatment
10 Hepatorenal failure,

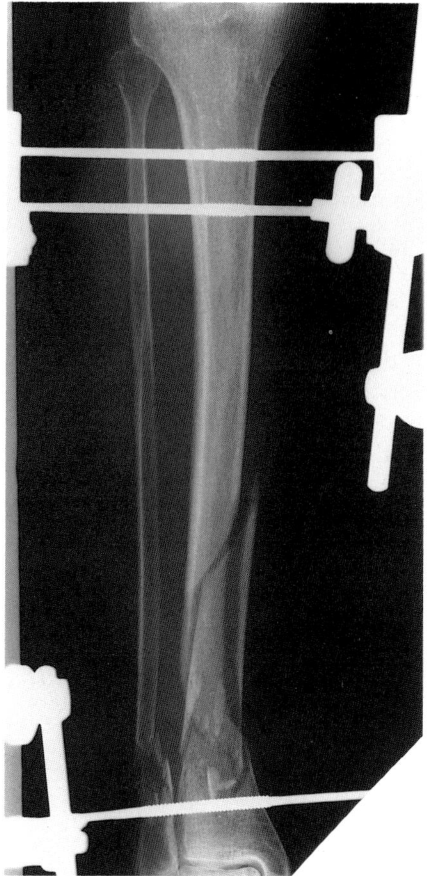

Fig. 2.19
X-ray showing comminuted fractures of the distal shafts of the right tibia and fibula in an elderly lady who developed an acutely ischaemic left leg.

and patients with:

11 Muscle necrosis due to the risk of acute renal failure from myoglobin release
12 Left atrial, left ventricular or aortic aneurysm thrombus due to the risk of further embolus

Prelysis clinical and radiological patient management

The causes of acute arterial ischaemia of the lower limb include:

1 Arterial thrombosis
2 Arterial embolus
3 Graft thrombosis
4 Thrombosis or embolus after an angiographic procedure

When a patient presents with critical lower limb ischaemia it is often difficult to distinguish between arterial thrombosis and embolus, and to assess the duration of the ischaemia, as most patients have a degree of atherosclerosis in their peripheral lower limb vessels. A history of intermittent claudication or rest pain suggests arterial thrombosis, whereas a history of recent myocardial infarction suggests arterial embolus. The presence of atrial fibrillation or mitral valve disease on clinical examination also suggests arterial embolus.

The majority of patients are therefore referred for urgent arteriography after Doppler arterial pressure measurements for assessment of their ankle-brachial systolic index, electrocardiography (ECG) to document the cardiac rate and rhythm, and echocardiography to look for left ventricular thrombus. A Doppler pressure below 40 mm Hg indicates critical ischaemia, but pressures of 0 mm Hg are often recorded in this type of patient.

It is important to obtain informed consent from the patient prior to the administration of a premedication, before arteriography. The following points should be made to a patient who may go on to receive intra-arterial thrombolytic therapy after the arteriography:

1 The condition of the leg is very serious and the aim of treatment is to prevent amputation
2 The limb salvage rate is about 75%
3 There is a risk of bleeding, requiring treatment in about 5% of cases
4 There is a risk of a stroke in about 1–2% of cases
5 The treatment is safer than surgery, which may also be required however
6 The treatment takes about 24 hours

Premedication with temazepam, 10–20 mg, orally, and morphine, 10 mg, orally, is advisable. The drugs can be given intravenously, but should not be administered intramuscularly in order to avoid an intramuscular haematoma.

Diagnostic arteriography is performed by retrograde catherization of the contralateral femoral artery (i.e. the femoral artery on the opposite side to the critically ischaemic leg) using a 4 or 5 Fr gauge pigtail catheter and nonionic contrast medium.

Ideally the arterial catheterization should be done using a thin-walled 19 gauge needle with only an anterior wall puncture of the common femoral artery. Puncture of the anterior and posterior walls of the artery with subsequent withdrawal of the tip of the needle into the lumen of the artery increases the risk of a groin haematoma, but may be unavoidable. A 4 or 5 Fr gauge pigtail catheter with multiple side holes near its tip can then be advanced over a 0.89 mm diameter (0.035 inch) Teflon-coated guidewire with a 3 mm J-shaped tip and positioned with its tip in the lower abdominal aorta below the renal arteries. The use of a 4 Fr catheter for diagnostic arteriography decreases the risk of a groin haematoma, if antegrade catheterization of the ipsilateral femoral artery is subsequently performed for placement of the arterial line. Lower limb peripheral arteriography should be performed with a low osmolar, nonionic, contrast medium at a strength of 300–370 mg I/ml, if conventional films are obtained or 150–200 mg I/ml, if a digital subtraction system is being used.

The arteriogram should demonstrate both the site and length of the occlusion and the run off vessels if they are patent, but is unlikely to be able to distinguish accurately between arterial thrombosis and embolus because the cut off appearances of both can look very similar (Figs 2.11 and 12a). The presence of multiple occlusions in several different arteries does however suggest arterial emboli. At this stage it is sensible to review the clinical findings and the radiological information with the vascular surgeons in the department of radiology so that a final decision regarding the patient's overall management can be made. If emergency surgery is required the diagnostic arteriography catheter is removed, before the patient goes to theatre, but if intra-arterial thrombolysis is necessary the catheter used to deliver it must be positioned correctly.

Occlusions that start in the common or external iliac artery, the common femoral artery or the proximal superficial femoral artery are best treated by positioning a 4 or 5 Fr gauge sidewinder No. 1 catheter with only an end hole across the aortic bifurcation, after forming it in the aortic arch. A sidewinder No. 1 catheter is easier to form in the aortic arch than a sidewinder No. 2 catheter and is also easier to manipulate in a narrow atheromatous aorta, but does not have the long tip of the sidewinder No. 2 catheter, which is useful for reaching the external iliac artery. The tip of a 4 or 5 Fr gauge sidewinder catheter is then positioned 2–3 cm into the common or external iliac artery occlusion (Fig. 2.20). The use of a 5 Fr rather than a 4 Fr catheter can be useful if the occlusion consists of firm rather than soft thrombus, because it is easier to advance. A 4 or 5 Fr gauge cobra catheter with only an end hole can also be used to cross the aortic bifurcation and does not need to be formed in the aortic arch. It is also easier to manipulate than a sidewinder catheter in tortuous iliac arteries, but has a tendency to enter the internal iliac artery. The tip of a 4 or 5 Fr gauge cobra catheter can then be positioned 2–3 cm into the common or proximal superficial femoral artery occlusion (Fig. 2.21).

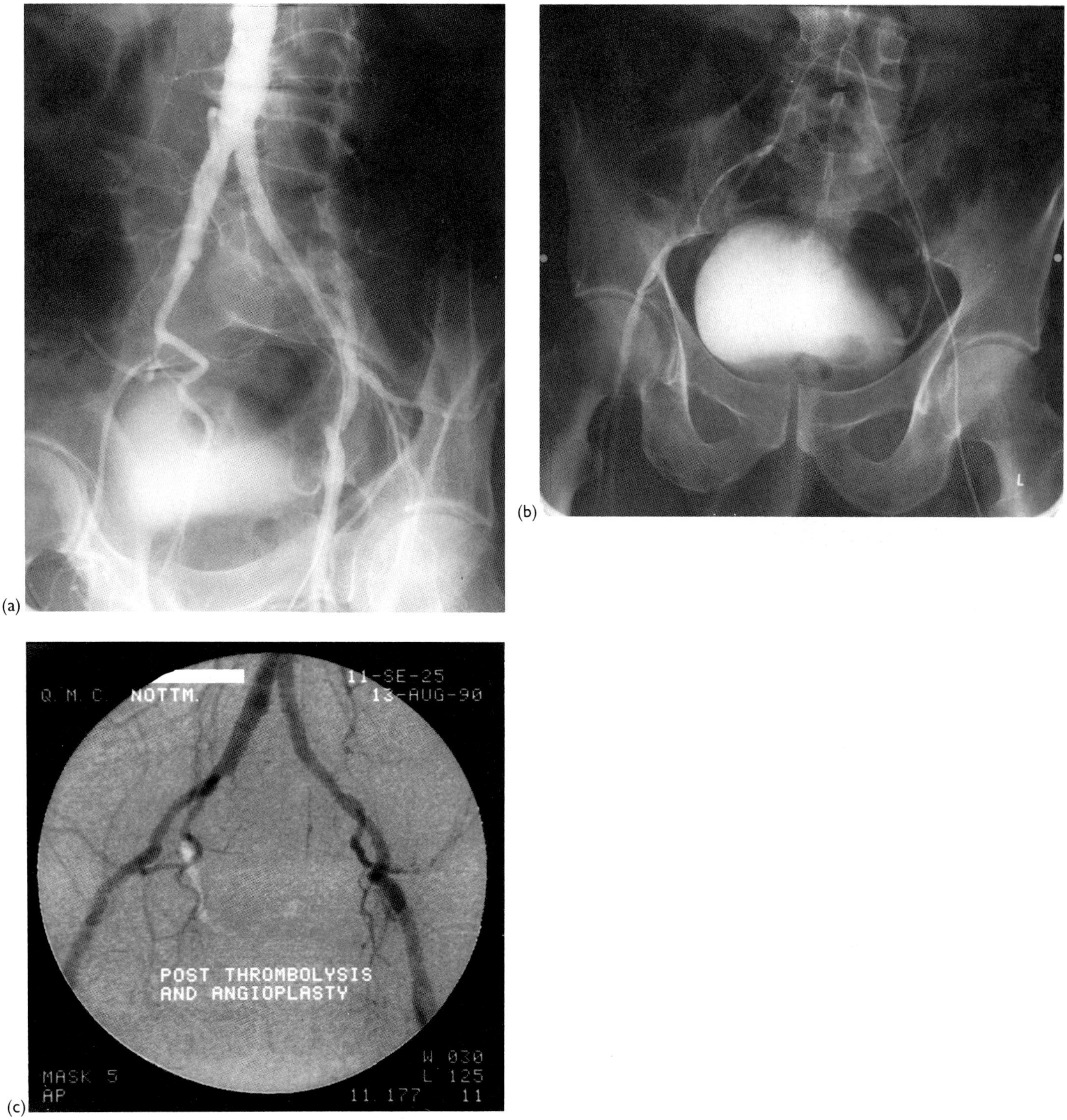

(a)

(b)

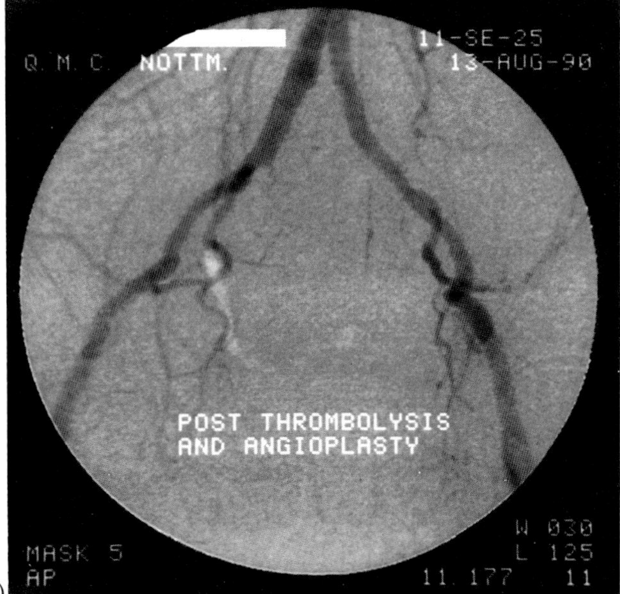

(c)

Fig. 2.20

(a–c) X-rays showing (a) right external iliac artery thrombosis extending into common femoral artery (oblique view); (b) during lysis with 5 Fr sidewinder catheter in the thrombus; (c) 5 years after successful lysis with angioplasty.

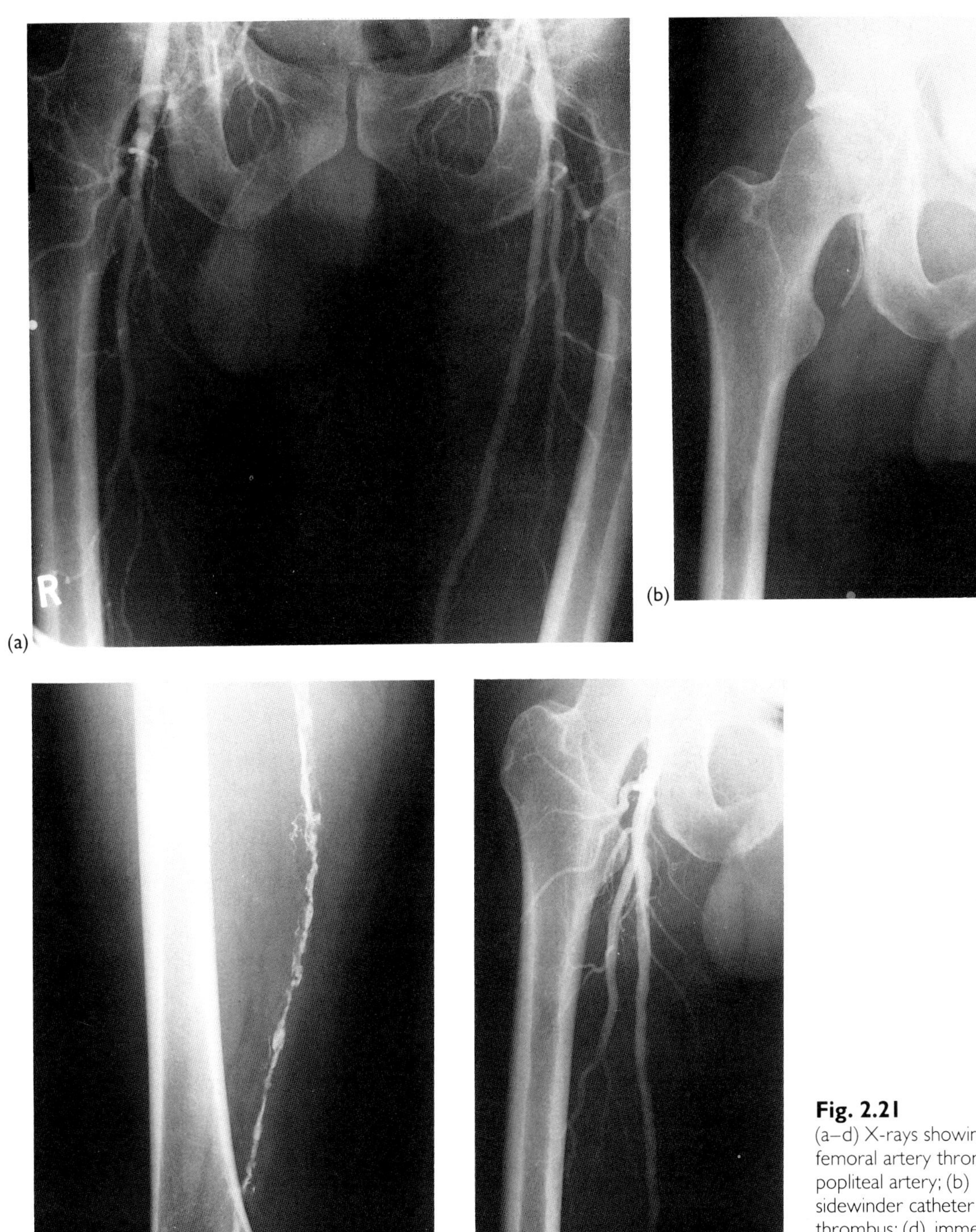

Fig. 2.21
(a–d) X-rays showing (a) right superficial femoral artery thrombosis extending into popliteal artery; (b) before lysis with 5 Fr sidewinder catheter in position; (c) in the thrombus; (d) immediately after successful lysis and angioplasty.

The tip of a 4 or 5 Fr gauge straight catheter can be positioned in occlusions in the distal superficial femoral or popliteal arteries after the use of a sidewinder and/or cobra catheter to cross the aortic bifurcation. However, tortuous iliac arteries may make it very difficult to advance any catheter this far (Fig. 2.22).

Therefore, occlusions that start in the distal superficial femoral, popliteal or tibial arteries are best treated by positioning a 4 or 5 Fr gauge straight catheter with several side holes as well as an end hole after antegrade catheterization of the ipsilateral common or superficial femoral artery (i.e. the femoral artery on the side of the critically ischaemic leg).

Again, this should be performed using a thin-

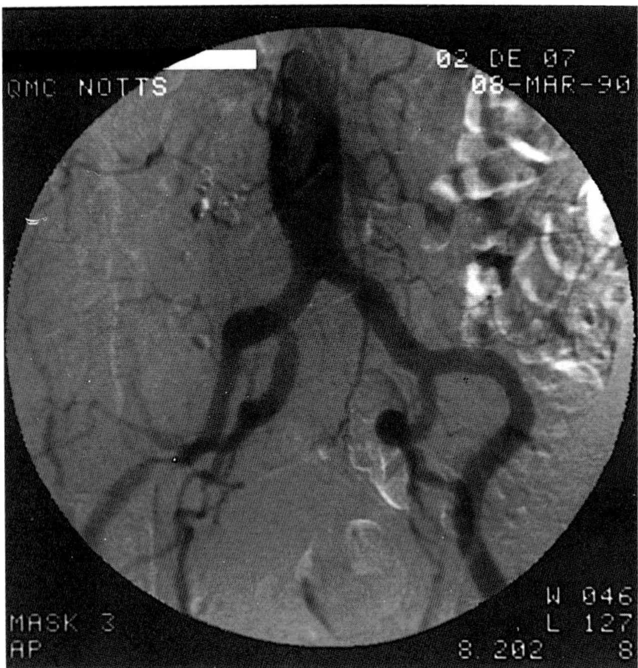

Fig. 2.22
X-ray showing very tortuous iliac arteries (and a stenosis in right external iliac artery).

walled 19 gauge needle and an anterior wall puncture of the common femoral artery. This additional arterial puncture increases the incidence of a groin haematoma, but is probably the optimal technique. Direct catheterization of the superficial femoral artery instead of the common femoral artery is useful in obese patients, where it may be difficult to get above the origin of the deep femoral artery, making it impossible to enter the origin of the superficial femoral artery. The tip of a 4 or 5 Fr gauge straight catheter is then positioned 2–3 cm into the distal superficial femoral or popliteal artery occlusion (Fig. 2.23). The use of a 4 Fr catheter can be useful if the arteries are narrow as this reduces the risk of pericatheter thrombosis.

The positioning of the tip of a catheter 2–3 cm into an occlusion is the ideal starting position for intra-arterial thrombolysis for occlusions up to 10–15 cm in length, whether the catheter has been introduced via the contralateral femoral artery and across the aortic bifurcation or into the ipsilateral femoral artery. However, for longer occlusions of 20–30 cm or more, the ideal starting position for intra-arterial thrombolysis is 10–15 cm or further into the occlusion, because the thrombolytic drug is usually able to break up thrombus up to this distance beyond the tip of the catheter (Earnshaw *et al.*, 1987a).

A hand injection of a few millilitres of nonionic contrast medium into the thrombus will confirm the position of the catheter tip and assess the patency of the distal vessels more accurately. The thrombus distal to the tip of the catheter should be probed gently with a guidewire to produce some fragmentation of the thrombus and so increase the surface area upon which the thrombolytic drug can act.

An intra-arterial injection of 5000 IU heparin into the thrombus should then be given in the department of radiology, before the thrombolytic regime begins. The catheter is secured on the skin surface under a clear plastic dressing so that the puncture site in the skin is visible and any bleeding is readily recognized.

Prelysis problems

Various difficulties may be encountered in positioning the catheter.

The difficult femoral arterial puncture

Retrograde or antegrade catheterization of the common femoral artery can sometimes be difficult if the femoral pulse is difficult to feel. Localization of the common femoral artery is made easier by:

1. Careful clinical palpation of a nonpulsatile cord (the pulseless femoral artery) in the groin in a thin patient
2. The use of fluoroscopy to visualize vascular calcification and bony landmarks
3. The use of duplex ultrasound to identify the vessels
4. The use of the roadmapping facility in a digital imaging system to demonstrate the artery

Antegrade catheterization of the common femoral artery is particularly difficult in obese patients and it is sometimes easier to catheterize the superficial femoral artery directly by directing the tip of the needle slightly medial to the maximum pulsation in the groin, which is produced by the common and deep femoral arteries.

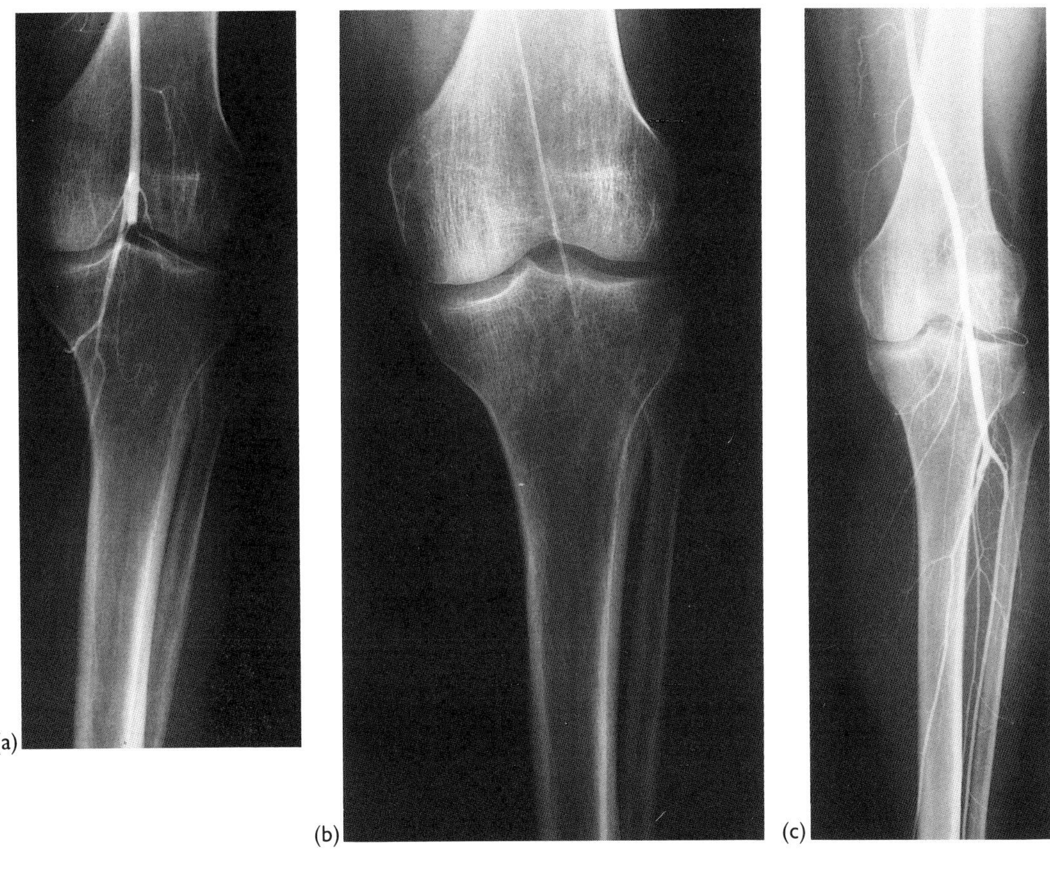

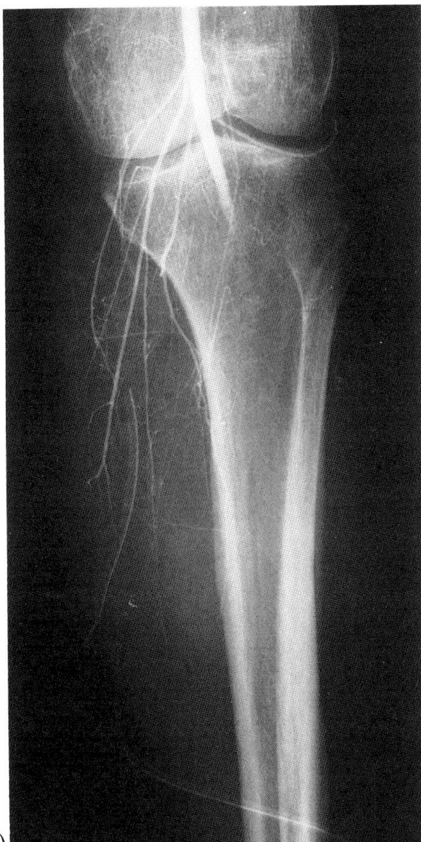

Fig. 2.23
(a–d) X-rays showing (a) left popliteal arterial embolus extending into tibial arteries; (b) before lysis with 4 Fr straight catheter in position; (c) immediately after successful lysis; (d) acute rethrombosis a few days later leading to amputation after unsuccessful surgery.

Manipulation of the catheter

Positioning of a catheter for either diagnostic arteriography or intra-arterial thrombolysis can be hindered by tortuous iliac arteries or atheromatous plaques in the iliac or femoral arteries. The use of gentle manipulation with either a curved catheter and a straight hydrophilic-coated guidewire or a straight catheter and an angled hydrophilic-coated guidewire will usually succeed.

The aortic bifurcation

Advancing a guidewire/catheter combination across the aortic bifurcation can sometimes be difficult, particularly if the iliac arteries are tortuous, because the guidewire/catheter combination tends to balloon up into the lower abdominal aorta if it does not advance. The use of a hydrophilic-coated guidewire or a stiffer Teflon-coated guidewire or even occasionally a 7 Fr catheter will usually prevent this happening.

Long occlusions

Positioning of the tip of the catheter distally within a long occlusion can be impeded by firm or hard thrombus. The use of hand injections of saline or contrast medium into the thrombus will often allow a guidewire/catheter combination to move further, but if this fails thrombolysis can be started from a proximal position within the long occlusion with subsequent advancement later, if lysis succeeds. The importance of a distal position within the thrombus is that by restoring the patency of the distal vessel, patent proximal collateral vessels can perfuse the ischaemic limb to some extent and an immediate bypass graft then becomes an additional potential treatment option.

Oozing from the skin puncture site

This can be an immediate problem in hypertensive patients, but generally stops after manual compression of the femoral artery in the groin. If the oozing continues it can be prevented by introducing a vascular sheath. The advantages of using a vascular sheath include easier catheter manipulation and less discomfort in the groin during subsequent catheter changes. The disadvantage of using a vascular sheath is the increased risk of perisheath thrombosis in a narrow vessel.

Arterial stenosis proximal to an occlusion

The presence of a significant arterial stenosis (greater than 50% of the lumen of the vessel), proximal to the occlusion to receive thrombolytic therapy, can produce a problem in that the presence of the catheter within the stenosis may occlude flow and cause thrombosis from the occlusion back to the stenosis. This can be prevented by initial angioplasty of the stenosis before the thrombolysis catheter is positioned within the occlusion or it can be treated by additional thrombolysis.

Two occlusions in the ischaemic leg

When there are two occlusions at different sites in the same limb it is usually not possible to tell which of them is recent and which is longstanding, unless there is previously documented clinical or radiological evidence of patent vessels. It is probably simplest to attempt to position a catheter in the more proximal occlusion to assess how soft or hard the thrombus feels. If the proximal occlusion is soft it will be easy to position the tip of the catheter as normal, but if the proximal occlusion is hard it will be difficult, although not impossible, to position the catheter and subsequent thrombolysis may be successful. If it is impossible to position a catheter in a proximal iliac occlusion, it may be feasible to place a catheter in the distal occlusion but this will require antegrade catheterization of a pulseless femoral artery. Reconstructive surgery is perhaps a better treatment option in this situation.

Tibial artery occlusion

Occlusions in the anterior tibial, posterior tibial or common peroneal arteries can be treated by positioning the tip of a 4 or 5 Fr catheter as normal, but because these vessels only have a diameter of 1–3 mm it is better to use either a 2 Fr catheter or a guidewire with an injection lumen in a coaxial system through either a 5 Fr vascular sheath or a 5 Fr catheter.

Aortic occlusion

An infrarenal occlusion of the lower abdominal aorta or occlusion of an aortic Dacron graft are probably best treated by reconstructive surgery, because positioning of a catheter for thrombolysis requires a high left brachial arterial puncture so that the catheter can reach the proximal end of the occlusion via the left axillary and subclavian arteries and the descending thoracic and upper abdominal aorta. In this situation there is obviously a risk of pericatheter thrombus embolizing up into the left vertebral artery on withdrawal of the catheter.

Extra-anatomic graft occlusion

Occlusion of either an axillofemoral or a femorofemoral cross-over Dacron graft can be treated by positioning a 4 or 5 Fr catheter within the graft following a direct puncture of it. The catheter can then be manipulated both ways to treat the proximal and distal thrombus.

THROMBOLYSIS MANAGEMENT

After the catheter has been positioned in the thrombus, the patient with a critically ischaemic leg should leave the department of radiology as soon as possible and either return to a specialist vascular ward or go to a bed in the intensive care unit. Intra-arterial thrombolysis can then be started with regular clinical (both medical and nursing), haematological (see section on haematological management) and radiological assessment of the patient, who is confined to bed. The following observations should be made during the infusion:

1 Blood pressure, pulse rate and puncture site check in the groin at 15 min intervals for the first hour after the procedure.
2 Then, blood pressure, pulse rate and puncture site check in the groin at 1 hourly intervals up to 4 hours after the procedure.
3 Then, blood pressure, pulse rate and puncture site check in the groin at 4 hourly intervals during the infusion.
4 Temperature measurement at 4 hourly intervals during the infusion.
5 Clinical assessment and Doppler pressure measurements at 4 hourly intervals during the infusion.
6 Blood clotting studies at 24 hourly intervals during the infusion.
7 Portable or departmental arteriograms initially at about 4 hours into the infusion, then 8 hours later (i.e. at 12 hours into the infusion) and at 12 hourly intervals thereafter until either successful lysis occurs or the treatment fails (Walker and Giddings, 1985).

Additional arteriography should also be performed if there is clinical deterioration in the leg during the infusion.

There are three fibrinolytic drugs that can be used for intra-arterial thrombolytic therapy—streptokinase, urokinase and tissue plasminogen activator. Suitable infusion regimes for intra-arterial thrombolytic treatment for each drug are:

1 Streptokinase infusion
   (a) Streptokinase, 5000 units/hour
   (b) Heparin 250 units/hour
   (c) Normal saline solution 15 ml/hour
2 Urokinase infusion
   (a) Urokinase 50000 units/hour
   (b) Heparin 1000 units/hour
   (c) Normal saline solution 50 ml/hour
3 Tissue plasminogen activator (tPA) infusion
   (a) tPA 0.5 mg/hour
   (b) No heparin
   (c) Normal saline solution 15 ml/hour

The only method of assessing progress of lysis is by arteriography, which can be done either as a portable film on the ward or as a series of films in the department. The advantage of portable films is that the patient does not have to return to the radiology department several times a day or night whilst the advantage of departmental films is better image quality.

A portable single shot arteriogram is easily performed in the bed, after a control film of the appropriate area has been obtained by injecting warm nonionic contrast medium by hand at a strength of 300 mg I/ml using a volume of 10 ml for the popliteal segment, 15 ml for the femoral segment or 20 ml for the iliac segment depending upon the position of the catheter tip.

The 4 hour arteriogram film will hopefully show some lysis of the thrombus, but the infusion should continue whether it does or does not for a further 8 hours. The 12 hour arteriogram film may show complete lysis of the thrombus, partial lysis of the thrombus or no lysis of the thrombus. Partial lysis of the thrombus is most frequently seen on the 12 hour arteriogram film and so the infusion should continue for a further 12 hours after repositioning the catheter usually by withdrawing it about 5–10 cm into further thrombus. Complete lysis of the thrombus at this stage indicates that thrombolysis has been successful and that the catheter can be removed, unless the arteriogram reveals a residual stenosis that requires angioplasty. No lysis of the thrombus at this stage indicates that thrombolysis has not yet been successful. If there has been no deterioration in the clinical state of the critically ischaemic leg, the infusion can be continued for a further 12 hours, but if the leg is deteriorating the infusion should be stopped and the catheter removed.

The 24 hour arteriogram film may show complete lysis of the thrombus, partial lysis of the thrombus or no lysis of the thrombus. Complete lysis of the thrombus is more likely to be seen on the 24 hour arteriogram film, which may show:

1 A residual arterial or graft stenosis, which is the cause of the occlusion.
2 Ectatic arteries with no underlying residual stenosis indicating *in situ* thrombosis due to the slow flow as the probable cause of the occlusion.
3 Normal arteries with no underlying residual stenosis indicating embolus as the probable cause of the occlusion.

If there is no residual stenosis the catheter should be withdrawn into the patent proximal arteries and the infusion continued for a further 2 hours to remove any pericatheter thrombosis before the catheter is finally removed. If there is a residual stenosis, angioplasty should be performed whilst the arterial access is available (Fig. 2.24).

No lysis of the thrombus on the 24 hour arteriogram film indicates that thrombolysis has failed, because there has been no lysis on two successive films at 12 hours, and the catheter should be removed. Further partial lysis of the thrombus indicates that thrombolysis is working and so the infusion should continue for a further 12 hours with repositioning of the catheter as necessary. The thrombolysis can continue for up to 72 hours as long as there is arteriographic evidence of continuing lysis, no clinical deterioration of the limb and no bleeding complications. In fact the longest infusion time that has resulted in successful limb salvage in our local experience was 80 hours!

Thrombolysis problems

There should now be complete lysis of the intra-arterial thrombus, but various problems may be encountered.

Groin haematoma

This develops during thrombolysis in a significant number of patients, particularly if they are hypertensive, but can often be controlled with the use of a compression bandage around the pelvis and upper thighs incorporating a 500 ml bag of any infusion fluid. Occasionally a large groin haematoma, which cannot be controlled, results in the thrombolytic treatment being abandoned, particularly if there has not been much lysis of thrombus.

Catheter removal

Accidental removal of the infusion catheter by the nursing or medical staff or deliberate removal of the infusion catheter by a confused and restless patient is particularly annoying to the radiologist who has positioned it, but unfortunately this does occur occasionally. The catheter must be replaced, unless the patient is particularly uncooperative, and this is easiest if there is a vascular sheath still in position.

Kinking and breakage of the catheter

Both kinking of the catheter, which may prevent the fibrinolytic drug being infused through the catheter, and breakage of the catheter, which allows leakage of the fibrinolytic drug onto the skin as well as back bleeding, are more likely to occur with 2 Fr catheters than 5 Fr catheters. Kinking can occur anywhere between the skin puncture site and the hub of the catheter, whereas catheter fracture usually occurs at the hub end of the catheter where the two join. Some of the new plastics are resistant to kinking and most of the catheters now have a strengthening collar at the catheter/hub junction to prevent fracture. A kinked or broken catheter has to be replaced.

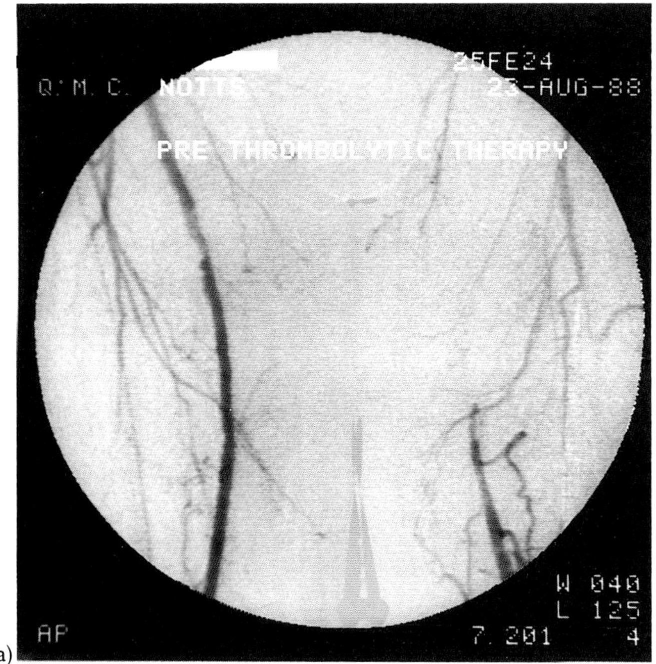

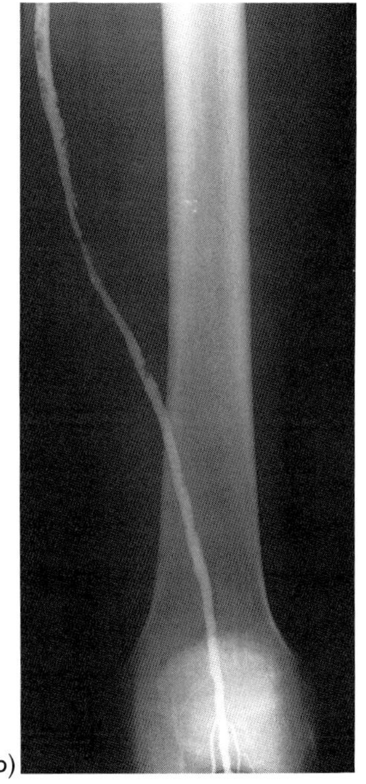

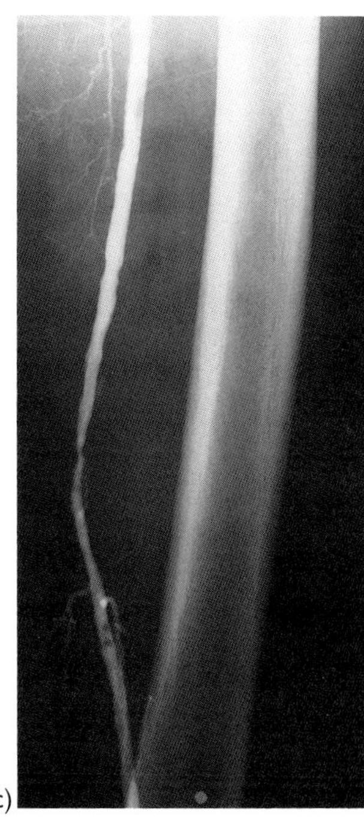

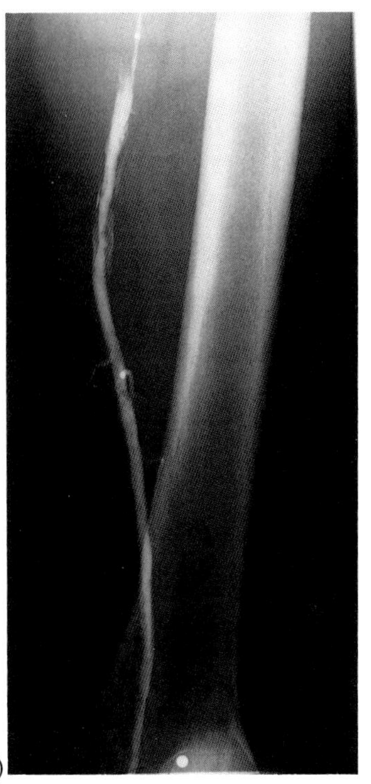

Fig. 2.24
(a–d) X-rays showing (a) left superficial femoral artery thrombosis; (b) lysis of the thrombus; (c) after successful lysis a residual stenosis is present; (d) after successful angioplasty.

Pericatheter thrombosis

This develops during thrombolysis and is more likely to occur in narrower arteries in which the pericatheter blood flow is slow. It can be detected clinically when, despite successful lysis of the thrombus, none of the distal pulses have become palpable and there has been no improvement in the perfusion of the leg. However, it is more likely to be demonstrated by arteriography. It is treated simply by withdrawing the tip of the catheter into the pericatheter thrombus and continuing with the thrombolytic treatment for about 2 hours (Fig. 2.25).

Distal embolization of thrombus

This also develops during thrombolysis and is less likely to occur in arteries in which the distal thrombus is lysed first. It can be detected clinically when, after successful lysis of the thrombus and the return of the distal pulses, there is a deterioration in the perfusion of the leg and loss of the distal pulses. It is better demonstrated by arteriography because it can occur before the thrombus is completely cleared. It is treated by advancing the tip of the catheter into the embolic thrombus and continuing with thrombolytic treatment for at least 4 hours (Fig. 2.26).

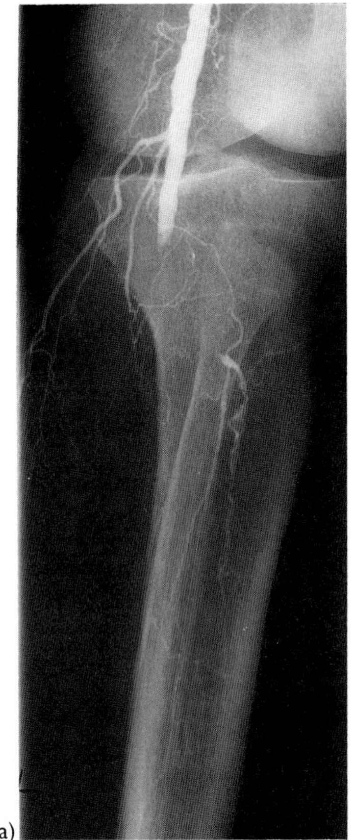
(a)

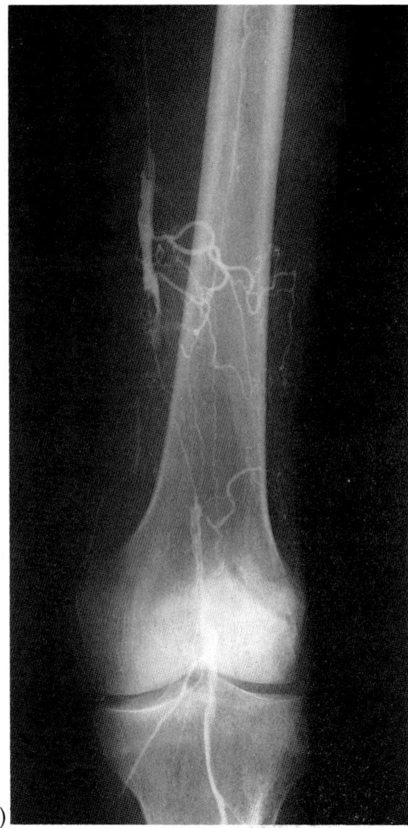
(b)

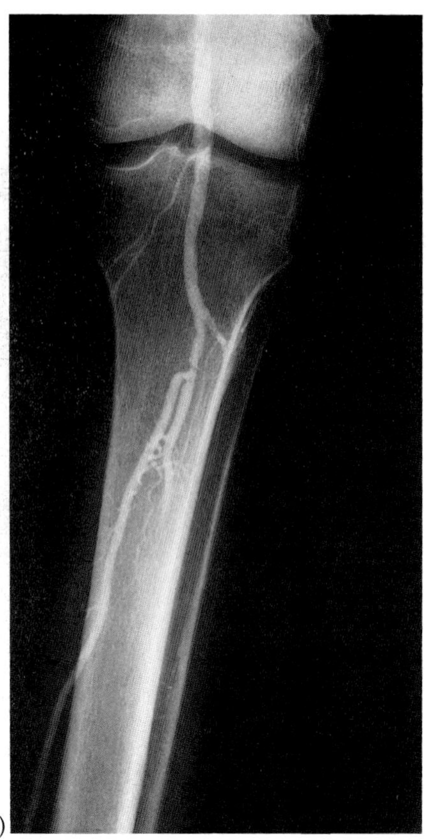
(c)

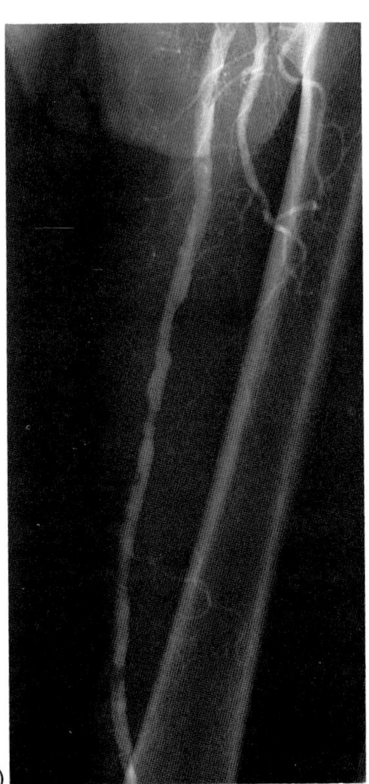
(d)

Fig. 2.25
(a–d) X-rays showing (a) left popliteal arterial thrombosis extending into tibial arteries; (b) during lysis with pericatheter thrombosis around 4 Fr catheter; (c) one week after successful lysis in popliteal artery; (d) one week after successful lysis in distal superficial femoral artery.

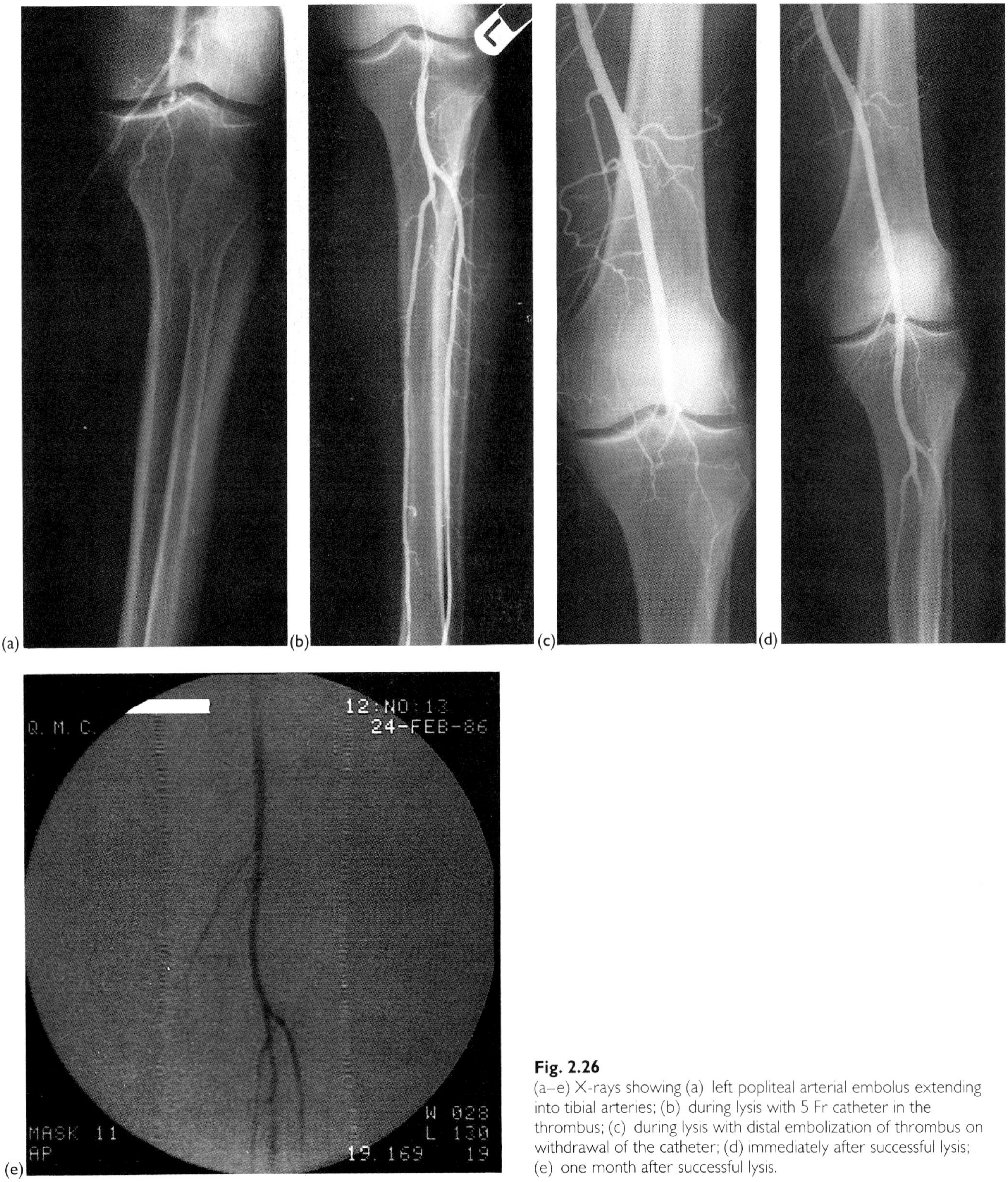

(a) (b) (c) (d)

(e)

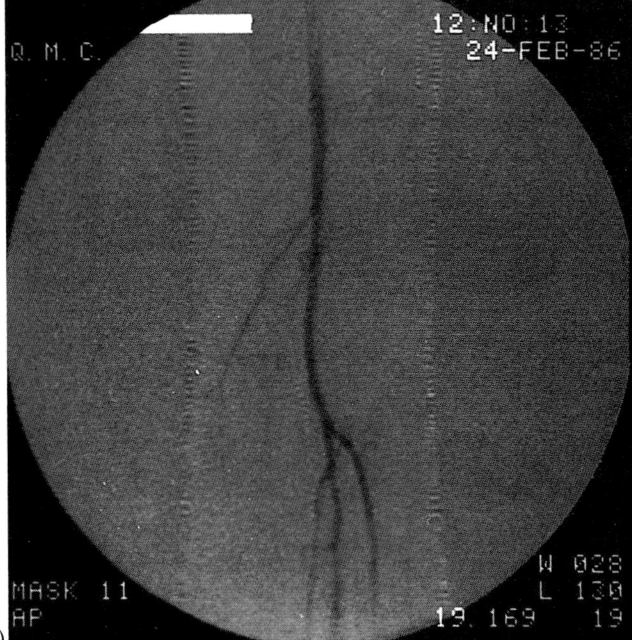

Fig. 2.26
(a–e) X-rays showing (a) left popliteal arterial embolus extending into tibial arteries; (b) during lysis with 5 Fr catheter in the thrombus; (c) during lysis with distal embolization of thrombus on withdrawal of the catheter; (d) immediately after successful lysis; (e) one month after successful lysis.

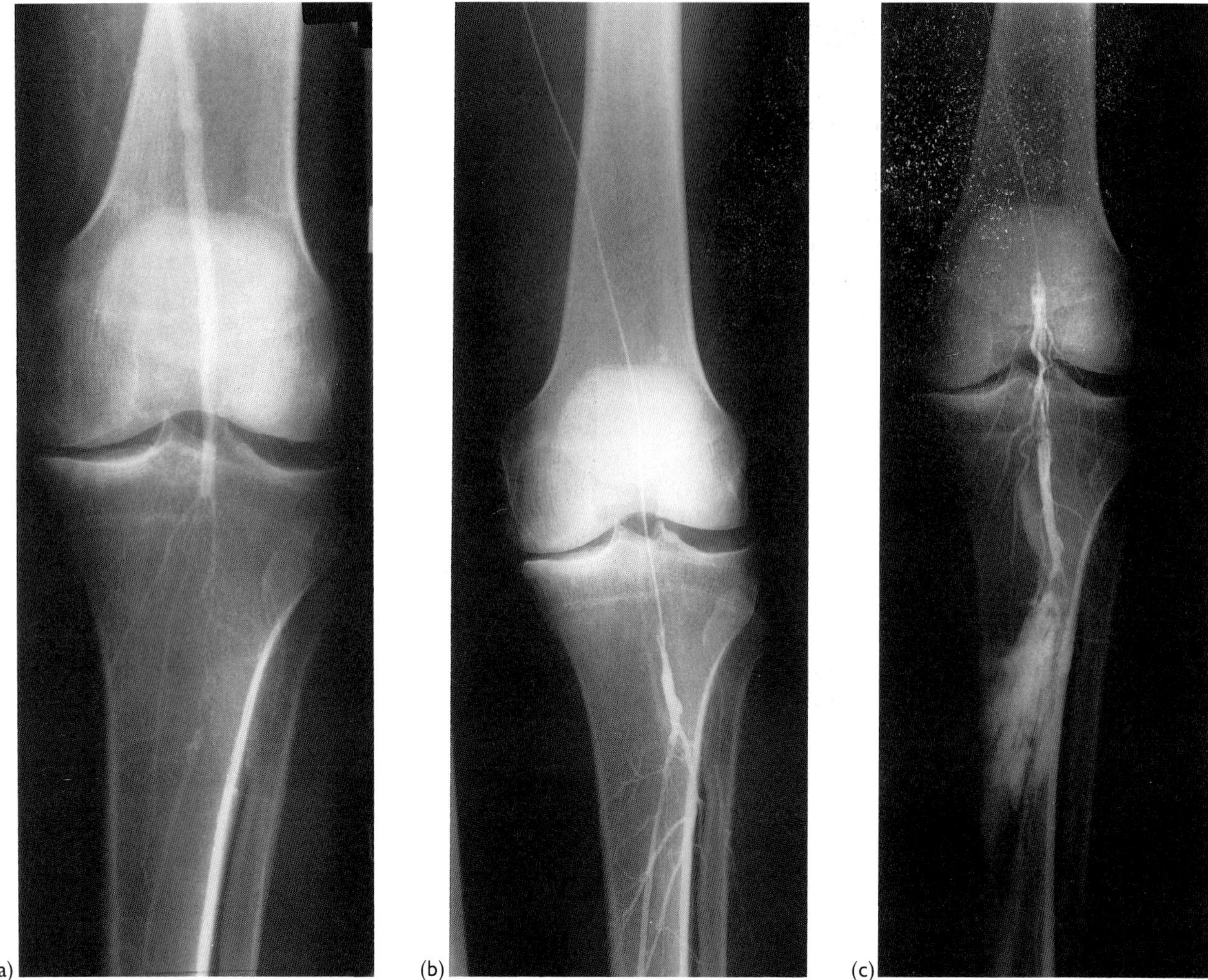

Fig. 2.27
(a–c) X-rays showing (a) left popliteal arterial embolus extending into tibial arteries; (b) during lysis with 4 Fr catheter in the thrombus; (c) during lysis with spontaneous extravasation of contrast medium indicating tibial artery perforation (limb salvage achieved).

Perforation of the artery

Perforation of the arterial wall with extravasation of contrast medium and blood into the perivascular soft tissues forming a haematoma can occasionally occur and may lead to the thrombolytic treatment being abandoned. It has been produced by the tip of the catheter becoming lodged in a small arterial branch of the popliteal artery around the knee joint (Earnshaw *et al.*, 1987b) with subsequent avulsion and has also occurred spontaneously from the tibial arteries (Fig. 2.27). It may also be produced by the tip of the guidewire or catheter during manipulation within the thrombus.

Gastrointestinal haemorrhage

The development of haematemesis, melaena or rectal bleeding is a serious complication that requires appropriate investigation and treatment. Thrombolytic therapy must be stopped immediately. Gastrointestinal bleeding is obviously more likely to occur in patients who have a potential bleeding site such as a peptic ulcer or tumour of the gastrointestinal tract. Severe epistaxis, haemoptysis or haematuria also indicate that the infusion must be stopped.

Retroperitoneal haemorrhage

Retroperitoneal haemorrhage is a serious complication which may occur spontaneously, be due to bleeding through the interstices of an aortic Dacron graft (Rabe *et al.*, 1982) or be related to puncture of the 'common femoral' artery above the inguinal ligament. Thrombolytic therapy must be stopped immediately and the patient transfused with whole blood, fresh frozen plasma and cryoprecipitate.

Cerebrovascular accident

A cerebrovascular accident is the most serious complication of thrombolytic therapy. It is most likely to occur at the site of a previous cerebral infarct, but may be due to either a spontaneous intracranial haemorrhage or a systemic embolus to the brain. The thrombolytic therapy must be stopped immediately and a computed tomographic (CT) scan arranged to assess both the cause and extent of this complication (Fig. 2.28).

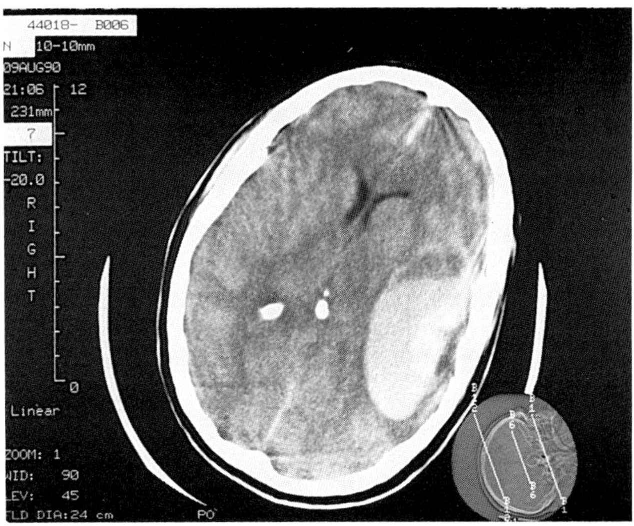

Fig. 2.28
Computed tomographic scan showing cerebral haemorrhage in left posterior parietal lobe.

Deterioration of the critically ischaemic leg

A critically ischaemic leg can occasionally deteriorate clinically during thrombolytic treatment despite successful lysis of the thrombus radiologically. Surgery in the form of fasciotomy or a by-pass graft to a distal patent segment of the arterial tree may result in limb salvage, but amputation may become necessary.

POSTLYSIS CLINICAL AND RADIOLOGICAL PATIENT MANAGEMENT

When complete lysis of the thrombus has been successfully achieved, the patient should return to the department of radiology for a completion arteriogram to assess the need for angioplasty of any underlying significant residual arterial stenosis, to look at the distal arteries for embolic thrombus and to check the proximal arteries for pericatheter thrombus, unless the portable arteriogram on the ward is of sufficient quality to answer these three questions.

If there is no residual arterial stenosis and no evidence of significant embolic or pericatheter thrombus, the infusion should be stopped and the catheter then removed. If there is either embolic or pericatheter thrombus, the catheter is repositioned and the infusion continued.

If there is a significant residual arterial stenosis, the patient needs to return to the department of radiology for a completion angioplasty before the catheter is removed. The infusion should be stopped and changed simply to normal saline, before the patient leaves the ward. The groin and catheter should then be cleaned with 1% chlorhexidine solution. The residual stenosis is crossed with a guidewire in the normal way and the infusion catheter can then be removed. An angioplasty balloon catheter on a 5 or 6 Fr shaft can then be positioned at the site of the stenosis. After an intra-arterial injection of 2000–5000 IU heparin, angioplasty is performed in the usual way. The presence of a vascular sheath in the groin probably makes the catheter manipulations more comfortable for the patient whose groin will undoubtedly be tender to palpation by this stage of the procedure.

Postangioplasty arteriography should then be

performed and if it indicates that this has been successful, the balloon catheter can be removed. It will be necessary to press on the groin at the site of the arterial puncture for at least 20–30 min to prevent a haematoma forming, and this is also likely to be uncomfortable for the patient, who can then return to the ward (Fig. 2.24).

Postlysis problems

There should now be a patent arterial system in a leg successfully treated by intra-arterial thrombolysis, but various problems may yet be encountered.

The difficult iliac artery stenosis

Crossing a residual iliac artery stenosis in an antegrade direction with a guidewire and catheter system can sometimes be difficult due to tortuous iliac arteries, atheromatous plaques in the iliac arteries or an eccentric residual stenosis in the iliac artery. Retrograde catheterization of the common femoral artery, which should now be palpable below the residual iliac artery stenosis, and crossing of the stenosis in a retrograde direction in the normal way will usually succeed (Fig. 2.29).

The difficult femoral or popliteal artery stenosis

Crossing a residual distal superficial femoral or popliteal artery stenosis with a guidewire and catheter system across the aortic bifurcation can sometimes also be difficult due to tortuous iliac arteries or the sheer distance from the arterial puncture site to the residual arterial stenosis, which can approach 100 cm. Antegrade catheterization of the common or superficial femoral artery, which is now patent above the distal superficial femoral or popliteal artery stenosis, and crossing the residual stenosis in an antegrade direction in the normal way will usually succeed.

Residual thrombus after angioplasty

The presence of a significant amount of residual arterial thrombus after angioplasty occasionally occurs, but is treated simply by repositioning the tip of the angioplasty catheter and repeating thrombolytic treatment for about 4 hours.

(a)
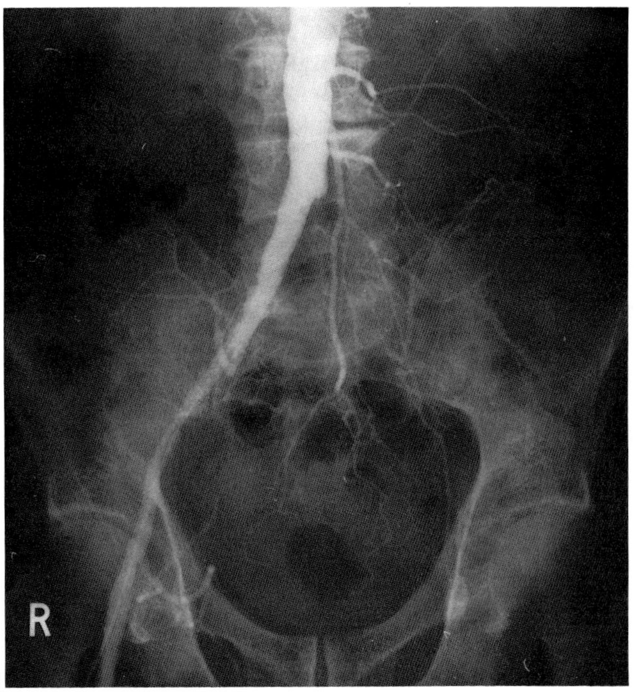

(b)
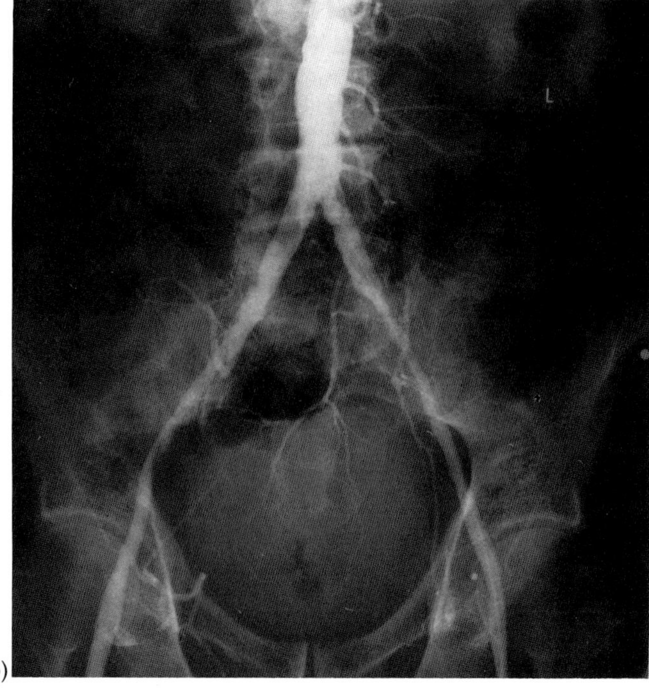

Fig. 2.29
(a,b) X-rays showing left iliac arterial thrombosis; after successful lysis a residual stenosis is present which required angioplasty.

Slow arterial flow

Slow blood flow in the arterial system, when there is good run off and no evidence of a residual stenosis, usually occurs in patients with poor cardiac output and ectatic lower limb arteries. This is likely to have been the cause of the original arterial occlusion and therefore obviously predisposes to rethrombosis. Improvement in the cardiac output and systemic anticoagulation may prevent this from happening.

Haematological management

The investigation of a patient with critical lower limb arterial ischaemia prior to starting intra-arterial thrombolysis includes:

1. Doppler arterial pressure measurements
2. Electrocardiography
3. Echocardiography
4. Diagnostic arteriography
5. Haematological laboratory investigations

There are a large number of potential haematological investigations which can be done, but before thrombolysis all patients should have venous blood taken for baseline measurements of the following tests:

1. Haemoglobin and full blood count
2. Urea, electrolytes and blood glucose
3. Blood group (and save serum for cross-matching)
4. Full clotting screen [international normalized ratio (INR) or prothrombin time (PT), activated partial thromboplastin time (APTT) and thrombin clotting time (TCT)]
5. Fibrinogen

Investigations such as plasminogen, α_2-antiplasmin, fibrinogen degradation products (FDPs), euglobin lysis time and the streptokinase tolerance test are interesting from a research point of view, but are unnecessary for the haematological management of uncomplicated thrombolysis.

During thrombolysis the haemoglobin, full blood count and clotting screen should be repeated on a daily basis. However, despite regular measurement of these haematological tests, the only evidence that the thrombus is actually lysing is obtained by repeat arteriography!

The haemoglobin, full blood count, clotting screen and fibrinogen tests should be repeated if there is active bleeding from any site and patients with either an unexplained tachycardia or a drop in blood pressure should be assumed to be bleeding occultly, unless another cause is apparent. Thrombolytic therapy must be stopped in the presence of active bleeding or if the fibrinogen level is less than 1.2 g/litre, as this is an indication of fibrinogen depletion in the circulation, and the patient resuscitated and treated with cryoprecipitate, fresh frozen plasma and whole blood transfusion.

After removal of the catheter following successful thrombolysis the patient should be started on an intravenous infusion of heparin at a dose of 1000 units/hour on returning to the ward. This level of anticoagulation should not be started if the thrombin clotting time is more than 60 sec. When the thrombin clotting time is less than 60 sec, heparin infusion can be started and the dose of heparin adjusted after 6 hours so that the activated partial thromboplastin time is maintained between two to three times of normal. Heparinization is maintained for 48 hours by measuring the activated partial thromboplastin time. Warfarin is then started, but the heparin infusion is continued until Warfarin control has been established. If there is a contraindication to the use of Warfarin, an antiplatelet drug such as aspirin or dipyridamole should be given for at least 6 months.

Anticoagulation is required after thrombolysis because of the risk of rethrombosis. This is caused by a rebound thrombotic state, which is partly due to platelet aggregation.

Fibrinolytic drugs

The clot dissolving or fibrinolytic system in the body is very complex, but can be simplified so that the various factors involved in the process of clot dissolution can be identified (Hessel and Kluft, 1986). These include the following and are shown in Fig. 2.30:

1. Activators of the fibrinolytic system, which trigger the conversion of circulating plasminogen to active plasmin.
2. The fibrin-dissolving enzyme plasmin, which is the key enzyme of the fibrinolytic system.
3. Blood clot containing fibrin, which is the natural substrate of the fibrinolytic system.

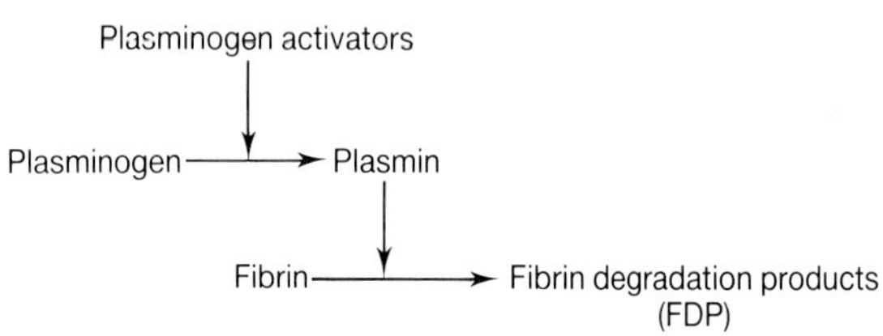

Fig. 2.30 Simplified diagram of the fibrinolytic system

Plasminogen activators

The fibrinolytic system is regulated by plasminogen activation, which can occur via three different routes:

1. Tissue plasminogen activator pathway
2. Factor XII-dependent plasminogen activator pathway
3. Factor XII-independent plasminogen activator pathway

Tissue plasminogen activator (tPA) is secreted by endothelial cells into the circulation by tPA release factors such as thrombin, bradykinin, acetylcholine, histamine and platelet-activating factor. Tissue plasminogen activator has a half-life of 2–5 min in the circulation and is inhibited by fast-acting plasminogen-activator inhibitor. The factor XII-dependent pathway produces kallikrein, which is a direct activator of plasminogen, and the factor XII-independent pathway produces urokinase, which is also a direct activator of plasminogen.

Plasminogen–plasmin system

Plasmin, which is produced by activation of plasminogen in the circulation, is an enzyme which has a broad nonspecific proteolytic activity against fibrin, fibrinogen, factor V and factor VIII. This is controlled in the circulation by a protease inhibitor, α_2-antiplasmin.

Fibrinogen–fibrin system

Fibrin, which is produced by the action of thrombin on the large pool of fibrinogen in the circulation, is the protein that forms the network, together with platelets, which traps the erythrocytes to produce intravascular thrombus. The fibrin network is broken up by the action of plasmin, with the production of fibrin degradation products in the circulation. Both plasminogen and tissue plasminogen activator have a high binding affinity for fibrin, which facilitates plasminogen activation on the surface of the molecule (fibrin specificity).

The fibrinolytic system can be activated by plasminogen activators such as streptokinase, urokinase and tPA to produce thrombolysis. The first generation agents, such as streptokinase and urokinase, interact with plasminogen in both the thrombus and the circulation, leading to systemic fibrinolysis. The second generation agents, such as recombinant tissue plasminogen activator (rtPA), acylated plasminogen streptokinase activator complex (APSAC) and single chain urokinase plasminogen activator (SCU-PA) are supposedly more fibrin specific and should therefore not produce significant systemic fibrinolysis, but this still occurs to a degree!

Streptokinase

Streptokinase is produced by Group C β-haemolytic streptococci and was discovered by Tillet and Garner (1933). It is a single chain polypeptide with a molecular weight of 47000 and has a half-life of 15–25 min, which initially forms a complex with plasminogen and this then activates both circulating and fibrin-bound plasminogen. Streptokinase is antigenic and some patients have antistreptococcal antibodies from a previous streptococcal infection. Streptokinase treatment should not be repeated in the same patient for at least a year because of this antigenicity. Streptokinase is used intra-arterially at a dose of 5000 (–10000) units/hour.

Urokinase

Urokinase is produced by human foetal kidney tissue culture cells and was isolated from urine by McFarlane and Pilling (1947). It is a double chain polypeptide with a molecular weight of 54000 and has a half-life of 12–16 min, which directly activates both circulating and fibrin-bound plasminogen. Urokinase is not antigenic. Urokinase is used intra-arterially at a dose of 50000 units/hour.

Tissue plasminogen activator

Tissue plasminogen activator is produced by the vascular endothelial cells and was originally derived from human uterine tissue and the Bowes melanoma cell, but is now produced by recombinant DNA gene technology after cloning of its gene by Pennica *et al.* (1983). Tissue plasminogen activator is a single chain glycoprotein with a molecular weight of 70 000 and has a half-life of 2–5 min, which binds to fibrin and then activates fibrin-bound plasminogen. Tissue plasminogen activator is not antigenic. Tissue plasminogen activator is used intra-arterially at a dose of 0.5 (–1) mg/hour.

Single chain urokinase plasminogen activator

Pro-urokinase is now also produced by recombinant DNA gene technology and is a single chain precursor form of urokinase, which activates fibrin-bound plasminogen, but does not bind to fibrin. It is not antigenic. Single chain urokinase plasminogen activator has not yet been used in acute lower limb ischaemia.

Anisoylated streptokinase plasminogen activator

This is a combination of streptokinase and plasminogen which binds to fibrin but is inactive until deacylation occurs. It then activates both circulating and fibrin bound plasminogen. It is antigenic. Anisoylated streptokinase plasminogen activator has been used intravenously at a dose of 5 mg tds.

Costs

The costs of the fibrinolytic drugs are:

1 Streptokinase in 100 000 unit vials costs £8 and therefore the cost of treatment for the majority of patients will be £16.
2 Tissue plasminogen activator in 20 mg vials costs £176 and therefore the cost of treatment for some of the patients will be £176, but for others it will be £352. Smaller vials may become available in the future.
3 Urokinase in 100 000 unit vials costs £60 and therefore the cost of treatment for patients will be about £1400.

COMPLICATIONS OF THROMBOLYSIS

Despite the use of low doses of the fibrinolytic drugs, haemorrhagic complications are still the major problem of intra-arterial thrombolytic therapy. The risk of a major haemorrhage (which is defined as a haemorrhage that produces hypotension and/or requires transfusion) is 5% and the risk of a minor haemorrhage is 15% (Berridge *et al.*, 1989). The complications can be classified as either localized to the ischaemic leg or systemic.

The complications localized to the ischaemic leg include:

1 Groin haematoma or haemorrhage from the arterial puncture site
2 Pericatheter thrombosis
3 Distal embolization of either lysing thrombus or forming pericatheter thrombus
4 Perforation of the arterial or graft wall with localized haemorrhage
5 Acute reocclusion after successful lysis of the thrombus

The management of most of these complications has already been discussed with a groin haematoma being controlled by an external compression bandage and both pericatheter thrombosis and distal embolization of thrombus being treated by further thrombolytic therapy. Perforation of the artery may prevent continuation of further thrombolytic therapy, but successful lysis has also occurred in this situation. Acute reocclusion of the occluded arterial segment within 30 days after successful lysis of the thrombus can be treated again by further intra-arterial thrombolytic therapy, preferably using tissue plasminogen activator. The use of streptokinase in this situation is contraindicated if streptokinase was used for the initial lysis, as there is a risk of anaphylaxis due to its antigenic nature. However, acute reocclusion may not require further treatment if there is no clinical deterioration in the patient's leg.

The systemic complications include:

1 Gastrointestinal haemorrhage
2 Retroperitoneal haemorrhage
3 Cerebrovascular accident
4 Acute renal failure
5 Hypersensitivity or allergic reactions to streptokinase

The management of most of these complications has also already been discussed with a cerebrovascular accident and either gastrointestinal or retroperitoneal haemorrhage requiring immediate cessation of treatment. Acute renal failure, due to the release of myoglobin from necrotic muscle and subsequent myoglobinuria, is preventable by good case selection. Patients with ischaemic necrosis of muscle should not be treated by thrombolytic therapy in the hope that a more distal amputation can be achieved. Allergic reactions to streptokinase range from pyrexia and skin rashes to serum sickness and acute anaphylaxis, and can be modified by the prior use of hydrocortisone, 100 mg, intravenously. Allergic reactions do not occur with tissue plasminogen activator and urokinase.

Results of thrombolysis

The majority of the published series on the treatment of acute arterial ischaemia of the lower limb with low dose intra-arterial thrombolytic therapy have described the use of streptokinase as the fibrinolytic drug, but over the last 5–10 years there has been increasing use of both urokinase and tissue plasminogen activator.

Using intra-arterial streptokinase successful lysis can be achieved in more than 60% of patients (with a range of 40–80%), resulting in limb salvage in 70–80% of cases after an adjunctive procedure (Earnshaw, 1991). Amputation is still necessary in 10–20% of patients and there is a mortality of less than 10%.

With intra-arterial urokinase successful lysis with limb salvage can be achieved in 70–80% of cases (Van Breda *et al.*, 1987), and with intra-arterial tissue plasminogen activator successful lysis with limb salvage can be achieved in 80–90% of cases (Graor *et al.*, 1986; Berridge *et al.*, 1990). Amputation is only necessary in 5–10% of patients, but there is still a mortality of about 10%.

The published results of intravenous thrombolytic therapy using anisoylated plasminogen activator streptokinase complex show successful lysis with limb salvage in 35% of patients (Earnshaw *et al.*, 1987c), which is similar to earlier publications. There was however a mortality rate of 30% with this treatment.

Strict criteria for successful thrombolysis (Lonsdale *et al.*, 1992a) have recently been suggested and are:

1. Radiological evidence of lysis with arterial recanalization at least as far as the next major collateral vessel.
2. An increase in the ankle–brachial systolic index of at least 0.2.
3. Limb salvage at 30 days.
4. No clinical evidence of rethrombosis within the first 30 days.
5. No intervention, other than angioplasty, at the level at which lysis was performed.

Early rethrombosis rates of 10–25% have been reported (Hess *et al.*, 1982; Lammer *et al.*, 1986), but the patency rates after successful thrombolysis are 80% at 1 year, 70% at 3 years and 60% at 5 years (Hess *et al.*, 1987; Lonsdale *et al.*, 1992b).

The key to success in thrombolysis lies in the multidisciplinary team approach of the radiologists, surgeons, haematologists, nurses and other technicians.

Recent advances in thrombolysis

The treatment of acute arterial ischaemia of the lower limb by thrombolysis would be improved by the following changes:

1. An increase in the rate of successful lysis
2. A decrease in the lysis time
3. A decrease in the incidence of haemorrhagic complications

These improvements should result in a higher limb salvage rate. Recent advances in thrombolysis management as well as future developments in thrombolysis are focussed on these ideals and therefore include:

1. New fibrinolytic drugs
2. New infusion regimes
3. New infusion catheters

An increase in the rate of successful lysis can be achieved by better patient selection, the use of surgery in combination with thrombolysis (e.g. in the patient with a thrombosed popliteal artery aneurysm, where thrombolysis is used to clear the run off vessels prior to definitive by-pass surgery) and the use of more effective fibrinolytic drugs. Tissue plasminogen activator, for example, has been shown to be superior to streptokinase in this respect in a small randomized trial of 40 patients (Berridge *et al.*, 1991). Single chain urokinase plasminogen activator has not as yet been used for thrombolysis in the peripheral vessels, but has been shown to have a synergistic effect in combination with tissue plasminogen activator. The use of prostaglandins and thromboxane inhibitors with current fibrinolytic drugs may also be helpful.

A decrease in the lysis time, which is dependent upon both the dose of the fibrinolytic drug in the infusion regime and the length of the occluding thrombus, can be produced with the use of more effective fibrinolytic drugs, new infusion regimes and better designed infusion catheters. Tissue plasminogen activator, for example, with a median time to lysis of 22 hours has been shown to be superior to streptokinase with a median time to lysis of 40 hours (Lonsdale *et al.*, 1992a). The mean infusion time for urokinase is also 22 hours (Van Breda *et al.*, 1987) and for streptokinase can be as low as 25 hours (Walker and Giddings, 1988). High-dose urokinase has a similar mean infusion time of 18 hours (McNamara, 1987), but both high-dose streptokinase (Hess *et al.*, 1987) or high-dose tissue plasminogen activator (Verstraete *et al.*, 1988) can have a mean infusion time of less then 6 hours.

Suitable high-dose infusion regimes for intra-arterial thrombolytic treatment for each drug are:

1 Streptokinase infusion
  (a) Streptokinase 20 000 units/hour (i.e. 1000 units/3 min)
  (b) No heparin
  (c) Normal saline solution 40 ml/hour
2 Urokinase infusion
  (a) Urokinase 240 000 units/hour (i.e. 4000 units/min)
  (b) Heparin 1000 units/hour
  (c) Normal saline solution 96 ml/hour
3 Tissue plasminogen activator (tPA) infusion
  (a) tPA 10 mg/hour
  (b) Heparin 400 units/hour
  (c) Normal saline solution 20 ml/hour

The technique of pulsed spray pharmacomechanical thrombolysis however is perhaps the most interesting recent development with a time to lysis of about 90 min! This technique, which was originally described by Bookstein *et al.* (1989) involves the use of small pulses of high-dose urokinase, which is sprayed throughout the thrombus via an infusion catheter, which has multiple side holes within the occlusion and its end hole occluded by a special tip-occluding guidewire. The pulses consist of 0.2 ml injections of a solution containing 25 000 units/ml urokinase every 15–20 sec using a 1 ml syringe. After 150 000 units of urokinase have been delivered in about 10 min, an additional 50 000 units are pulsed in over the next 15–20 min and once blood flow has been established 50 000 units/hour are infused until lysis is complete. The patients also receive 3000–5000 units of heparin at the beginning of the lysis and a further 3000–5000 units at the end of the lysis, prior to angioplasty. The use of mechanical rotational catheter systems may also prove helpful and are currently being evaluated.

REFERENCES

AMERY A, DELOOF W, VERMYLEN J, VERSTRAETE M (1970) Outcome of recent thromboembolic occlusions of limb arteries treated with streptokinase. *British Medical Journal* **iv**: 639–44.

BERRIDGE DC, MAKIN GS, HOPKINSON BR (1989) Local low dose intra-arterial thrombolytic therapy: the risk of stroke or major haemorrhage. *British Journal of Surgery* **76**: 1230–33.

BERRIDGE DC, GREGSON RHS, MAKIN GS, HOPKINSON BR (1990) Tissue plasminogen activator in peripheral arterial thrombolysis. *British Journal of Surgery* **77**: 179–82.

BERRIDGE DC, GREGSON RHS, HOPKINSON BR, MAKIN GS (1991) Randomized trial of intra-arterial recombinant tissue plasminogen activator, intravenous tissue plasminogen activator and intra-arterial streptokinase in peripheral arterial thrombolysis. *British Journal of Surgery* **78**: 988–95.

BLAISDELL FW, STEELE M, ALLEN RE (1978) Management of acute lower extremity arterial ischemia due to embolism and thrombosis. *Surgery* **84**: 822–31.

BOOKSTEIN JJ, FELLMETH B, ROBERTS A, VALJI K, DAVIS G, MACHADO T (1989) Pulsed spray pharmacomechanical thrombolysis: preliminary clinical results. *American Journal of Roentgenology* **152**: 1097–100.

COTTON LT, FLUTE PT, TSAPOGAS MJC (1962) Popliteal artery thrombosis treated by streptokinase. *Lancet* **ii**: 1081–3.

DALE WA (1984) Differential management of acute peripheral arterial ischaemia. *Journal of Vascular Surgery* **1**: 269–78.

DOTTER CT, ROSCH J, SEAMAN AJ (1974) Selective clot lysis with low-dose streptokinase. *Radiology* **111**: 31–7.

EARNSHAW JJ, GREGSON RHS, MAKIN GS, HOPKINSON BR (1987a) Early results of low dose intra-arterial streptokinase therapy in acute and subacute lower limb arterial ischaemia. *British Journal of Surgery* **74**: 504–7.

EARNSHAW JJ, GREGSON RHS, LOWE JS, HOPKINSON BR, MAKIN GS (1987b) Case report: Popliteal artery perforation during low dose intra-arterial streptokinase infusion. *Clinical Radiology* **38**: 321–3.

EARNSHAW JJ, WESTBY JC, MAKIN GS, HOPKINSON BR (1987c) Low dose intra-arterial streptokinase and acylated plasminogen–streptokinase activator complex: a retrospective review of two thrombolytic regimes in recent peripheral arterial ischaemia. *European Journal of Vascular Surgery* **1**: 151–8.

EARNSHAW JJ (1991) Thrombolytic therapy in the management of acute limb ischaemia. *British Journal of Surgery* **78**: 261–9.

FIESSINGER JN, AIACH M, VAYSSAIRAT M, JUILLET Y, CORMIER JM, HOUSSETT E (1978) Traitment thrombolytique des arteriopathies. *Annales De L'Anesthesiologie Francais* **19**: 739–45.

GRAOR RA, RISIUS B, LUCAS, FV, YOUNG JR, RUSCHHAUPT WF, BEVEN EG, GROSSBARD EB (1986) Thrombolysis with recombinant human tissue-type plasminogen activator in patients with peripheral artery and bypass graft occlusions. *Circulation* **74** (Suppl 1): 15–20.

HESS H, INGRISCH H, MIETASCHK A, RATH H (1982) Local low dose thrombolytic therapy of peripheral arterial occlusions. *New England Journal of Medicine* **307**: 1627–30.

HESS H, MIETASCHK A, BRUCKL R (1987) Peripheral arterial occlusions: a 6 year experience with local low dose thrombolytic therapy. *Radiology* **163**: 753–8.

HESSEL LW, KLUFT C (1986) Advances in clinical fibrinolysis. *Clinics in Haematology* **15**: 443–63.

LAMMER J, PILGER E, NEUMAYER K, SCHREYER H (1986) Intra-arterial fibrinolysis: long term results. *Radiology* **161**: 159–63.

LEVEEN HH, DIAZ CA (1972) Venous and arterial occlusive disease treated by enzymatic clot lysis. *Archives of Surgery* **105**: 927–35.

LONSDALE RJ, BERRIDGE DC, EARNSHAW JJ, HARRISON JD, GREGSON RHS, WENHAM PW, HOPKINSON BR, MAKIN GS (1992a) Recombinant tissue-type plasminogen activator is superior to streptokinase for local intra-arterial thrombolysis *British Journal of Surgery* **79**: 272–5.

LONSDALE RJ, WHITAKER SC, BERRIDGE DC, EARNSHAW JJ, GREGSON RHS, WENHAM PW, HOPKINSON BR, MAKIN GS (1992b) Long term results of peripheral thrombolysis. In press.

MCFARLANE RG, PILLING J (1947) Fibrinolytic activity of normal urine. *Nature* **4049**: 779.

MCNAMARA TO (1987) Role of thrombolysis in peripheral arterial occlusions. *American Journal of Medicine* **83** (Suppl 2A): 6–10.

MCNICOL GP, REID W, BAIN WH, DOUGLAS AS (1963) Treatment of peripheral arterial occlusion by streptokinase perfusion. *British Medical Journal* **i**: 1508–12.

MCPHAIL NV, FRATESI SJ, BARBER GG, SCOBIE TK (1983) Management of acute thromboembolic limb ischaemia. *Surgery* **93**: 381–5.

PENNICA D, HOLMES WE, KOHR WJ, HARKINS RN, VEHAR GA, WARD CA, BENNET WF, YELVERTON E, SEEBURG PH, HEYNEKER HL, GOEDDEL DV, COLLEN D (1983) Cloning and expression of human tissue type plasminogen activator cDNA in *E. coli. Nature* **301**: 214–21.

RABE FE, BECKER GJ, RICHMOND BD, YUNE HY, HOLDEN RW, DILLEY RS, KLATTE EC (1982) Contrast extravasation through Dacron grafts: a sequela of low dose streptokinase therapy. *American Journal of Roentgerology* **138**: 917–20.

REPORT OF A NATIONAL INSTITUTES OF HEALTH CONCENSUS DEVELOPMENT CONFERENCE (1980) Thrombolytic therapy in thrombosis. *Annals of Internal Medicine* **93**: 141–4.

TAWES RL, HARRIS EJ, BROWN WH, SHOOR PM, ZIMMERMAN JJ, SYDORAK GR, BEARE JP, SCRIBNER RG, FOGARTY TJ (1985) Arterial thromboembolism. A 20 year perspective. *Archives of Surgery* **120**: 595–8.

TILLETT WS, GARNER RL (1933) The fibrinolytic activity of haemolytic streptococci. *Journal of Experimental Medicine* **58**: 485–502.

VAN BREDA A, KATZEN BT, DEUTSCH AS (1987) Urokinase versus streptokinase in local thrombolysis. *Radiology* **165**: 109–11.

VERSTRAETE M, HESS H, MAHLER F, MIETASCHK A, ROTH FJ, SCHNEIDER E, BAERT AL, VERHAEGHE R (1988) Femoro-popliteal artery thrombolysis with intra-arterial infusion of recombinant tissue-type plasminogen activator: a report of a pilot trial. *European Journal of Vascular Surgery* **2**: 155–9.

WALKER WJ, GIDDINGS AEB (1985) Low dose intra-arterial streptokinase: Benefit versus risk. *Clinical Radiology* **36**: 345–54.

CHAPTER 3

Embolization

Anne P. Hemingway

Indications 62

Materials and Equipment 62

Technique 67

Premedication and anaesthesia 67

Preliminary angiography 67

The choice of embolic agent 68

Specific indications 68

Systemic arteriovenous malformation 68

Trauma 74

Tumours 75

Aftercare and follow-up 77

Complications 77

References 79

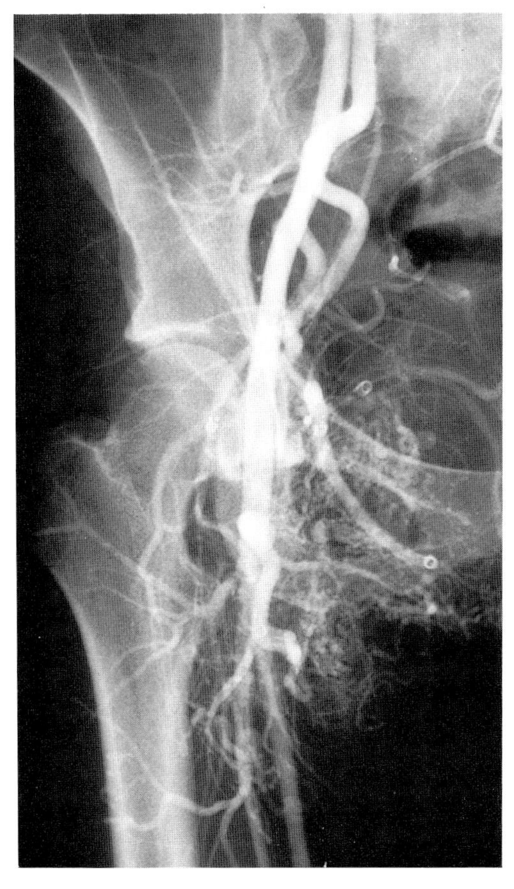

Therapeutic embolization refers to those procedures which involve the deliberate occlusion of blood vessels and vascular spaces by the injection, under fluoroscopic control, of embolic materials through selectively placed catheters.

Indications

Embolization can be performed in a wide range of conditions either as a planned or elective procedure or in the emergency situation. In many cases and conditions a single procedure suffices but in some conditions, most notably arteriovenous malformations (AVMs), repeated procedures may be necessary to gain control of a lesion.

Embolization can be performed as the definitive form of treatment in nonmalignant lesions or preoperatively or palliatively in benign and malignant conditions.

The range of indications for embolization in the peripheral vascular system is listed in Table 3.1. These are discussed in more detail below following the more general discussion on features pertinent to all procedures.

Materials and equipment

Embolization procedures should be regarded in the same light as other complex surgical procedures. They require skilled personnel and the use of high-quality equipment if the outcome is to be successful and free of complication.

Personnel

Radiologists undertaking embolization procedures must be skilled and experienced angiographers with a sound knowledge of the techniques and conditions being treated. He/she should preferably be working in a centre where these and other vascular procedures are performed frequently and should maintain regular contact with other embolists in order to be aware of new developments. Inevitably, from time to time, a relatively inexperienced vascular radiologist will be confronted with an emergency (e.g. exsanguination). If help is not at hand and it is impossible to transport the patient to another centre then it is acceptable to do whatever is practically possible to stabilize the patient's condition.

There are good medical and educational, as well as economic, arguments for the establishment of a small number of centres nationwide which specialize in the more complex embolization procedures (e.g. systemic and pulmonary AVMs) where the expertise exists, where training can be undertaken and where the inevitable large range of often expensive consumables needed can be concentrated. These centres would exist where the appropriate clinical backup in terms of vascular surgeons, plastic surgeons, haematologists, dermatologists etc. are also present. The vast majority of radiologists experienced in the more complex procedures are more than happy for the referring radiologists to accompany the patient and be actively involved in the procedure and subsequent management of the patient.

Radiographic equipment

Ideally the highest possible quality imaging equipment is required. It is essential that images can be recorded either on film or digitally at a very fast rate (up to 6 frames per sec) as many lesions exhibit very

Arteriovenous malformations (AVMs)	Definitive Preoperative Palliative
Trauma	Haemorrhage Arteriovenous fistulae
Tumours	Definitive (benign lesions) Preoperative – to reduce size/vascularity (benign/malignant) Palliative – to reduce symptoms (benign/malignant)

Table 3.1 *Indications for embolization in the peripheral vascular system*

rapid flow and arteriovenous shunting. Images need to be rapidly available for constant review throughout the procedure. High-quality fluoroscopy is also necessary particularly during the actual embolization when apparently very minor changes in haemodynamics must be recognized immediately and the course through the vessels of the embolic agent (e.g. coils) monitored closely.

Modern digital subtraction angiographic (DSA) equipment offers the best possible solution. Rapid image acquisition with 'real-time' subtraction, efficient image storage and retrieval, image manipulation, roadmapping and speed of examination enhance the safety of the procedure. The added advantage of DSA is the ability to use low concentrations and small volumes of contrast media to obtain diagnostic images. This is critically important in patients, for example, with large AVMs where many angiographic studies are needed to delineate the lesion accurately.

'Disposable' equipment – catheters, wires etc.

In order to provide a comprehensive embolization service the radiologist needs to have access to a large amount of disposable equipment to cover most eventualities, expected and unforeseen. It is unprofessional to commence an elective embolization procedure and be unable to complete it for lack of a standard catheter or wire, or sufficient or appropriate embolic materials. From time to time a totally unforeseen situation may arise and a procedure has to be terminated while, for example, a new or custom-made catheter or coil is ordered.

CATHETERS (Berenstein and Krichett, 1979; Chuang, 1990; Allison and Machan, 1992)

The author will not enter into the debate as to whether 5 or 7 Fr catheters are best. An angiographer should use whichever one he/she is most comfortable with and with which the highest number of successful results is achieved. Whatever system is adopted it is essential to ensure that all the elements, for example, guidewires, coils etc. are compatible (e.g. some 5 Fr catheters will only take 0.035 inch wires and are therefore unsuitable for standard 0.038 inch coils, but will accept 0.035 inch or smaller coils). It is essential to chose a catheter with good torque control and radioopacity. The most useful catheters in the author's experience are the:

Headhunter (Femorocerebral A)
Sidewinder (Femorocerebral B) I, II, III
Cobra (Femorovisceral) I, II, III

Coaxial 3 Fr catheter systems have proved to be invaluable adjuncts to conventional catheters in embolization procedures (e.g. the Tracker catheter etc.). Its small size, flexibility, torquability and ability to deliver coils have rendered such catheters an essential part of the embolist's armamentarium.

NB, for embolization, catheters should have an end hole only and NO side holes, as particulate emboli may become lodged in the side hole, be impossible to dislodge with a wire, but will then be 'wiped' off into a vulnerable vascular bed on withdrawing the catheter.

GUIDEWIRES

It is essential to have a large selection of shapes, lengths and sizes of guidewires available. The 0.035 inch wire is compatible with most widely available 5 and 7 Fr catheter systems. Suggested wires include:

Standard length 145 cm wires:	Straight, 3 cm floppy tip Newton cerebral, 15 cm floppy tip J-curves 1.5, 3, 7.5 and 15 mm Heavy duty/Rosen wires
Exchange length 180 and 260 cm:	Straight Newton cerebral J-curve 3 mm Heavy duty

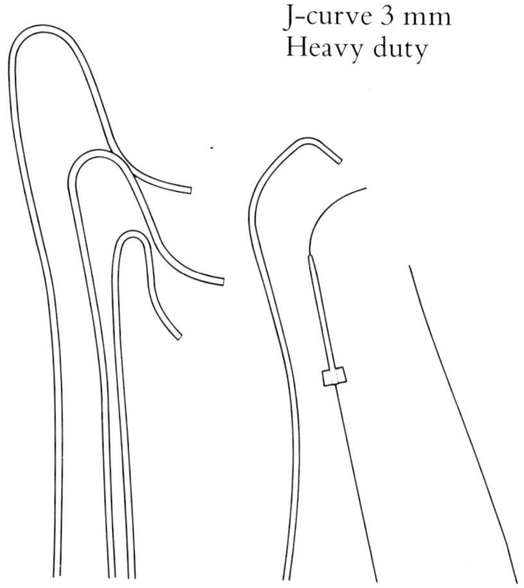

Fig. 3.1 Commonly used angiographic catheters and wires.
From left to right, Sidewinders III, II, I, Cobra I, 15 mm J-wire and Newton cerebral 15 mm floppy tip wire.

Terumo, straight and curved tip, standard length, extra long and super-stiff varieties are available and useful. This wire, also known as the 'glide wire' or 'slime wire', has a hydrophilic coating and has at all times to be kept wet. It is remarkably atraumatic, and once the operator has mastered the knack of handling the wire (it is *very* slippery) it will enable areas inaccessible to other wires to be catheterized.

EMBOLIC MATERIALS (Novak, 1990b; Hemingway, 1986)

There is a large variety and range of embolic agents available and the choice of which one or ones to use depends on the site, size and nature of the lesion being treated and whether or not temporary or permanent occlusion is required. Some currently available agents are listed in Table 3.2. Those marked with an asterisk represent the most commonly used agents in the UK and will be discussed in more detail. It is beyond the scope of this text to discuss all the agents and the reader is referred to larger reference texts.

The ideal embolic agent would have the following properties:

1 Ready availability
2 Wide range of sizes
3 Sterile
4 Nontoxic
5 Radioopaque
6 Reliable and consistent effects
7 Low cost

The ideal agent which can be used in any and every situation does not exist, hence the wide variety of agents available.

Embolic agent	Occlusion time
Natural products	
Autologous blood clot	Temporary
Autologous muscle fragments	Semipermanent
Autologous fat	Semipermanent
Lyophilized human dura mater	Semipermanent
Microfibrillar collagen (Avitene)	Semipermanent
Collagen fibrils (Tachotop)	Semipermanent
Particulate emboli (synthetic)	
*Sterile absorbable gelatin sponge Sterispon/Gelfoam/Spongostan	Temporary
*Polyvinyl alcohol (Ivalon) sponge	Permanent
*Contour	Permanent
*Gianturco coils	Permanent
*Detachable ballons	Permanent
Variable size silicon/glass spheres	Permanent
BISS (barium-impregnated silicon sogeres)	Permanent
Liquids	
*Hypertonic dextrose	Permanent
*Absolute alcohol	Permanent
Isobutyl-2-cyanoacrylate (Superglue)	Permanent
Boiling contrast medium/saline	Permanent
Ethibloc	Permanent
Lipiodol	
Other	
Electrocoagulation	Permanent
Laser therapy	Permanent
Percutaneous embolization	Permanent

*Most commonly used agents in the UK.

Table 3.2

It is of paramount importance to select an appropriate agent and of the size most likely to occlude those vessels being embolized without passing through if too small or lodging proximally if too large. Emboli delivered into the systemic circulation may pass through a high flow lesion and will, if particulate, become trapped in the pulmonary vascular bed. It is inadvisable for this to happen to any significant degree as it represents iatrogenic pulmonary embolism.

A particulate embolus, especially a coil, lodged too proximally will not only fail to treat the lesion adequately, but will also prevent further attempts at effective embolization. The underlying principle is to occlude a lesion from the centre outwards, thus obliterating the nidus or core and thus reducing the likelihood of recanalization and collateral formation.

It is inadvisable to use liquid emboli in areas where they may permeate through small vessels into adjacent, vulnerable areas, for example, internal iliac system when the blood supply to the sciatic nerve may be damaged, or when there is a significant risk of causing skin necrosis as in some vascular malformations.

Particulate emboli

Gelatin sponge (Gelfoam/Spongostan) A sterile absorbable gelatin sponge is a versatile and useful agent (Sterispson is currently not available in the UK). It comes packed in sterile sheets which can be cut into particles of the desired size. It is advisable to use sterile sharp, straight dry scissors to cut the foam and it is also worth putting on fresh, dry sterile gloves as the foam becomes difficult to handle once wet. Gelfoam powder is available in some countries and is very useful for embolizing small vessels. As the sponge is not radioopaque it is delivered suspended in contrast medium so that its passage through the vessels can be monitored. The sponge acts as a matrix for thrombus formation and platelet aggregation. Inflammatory changes may occur in the vessel wall but the occlusion is temporary. Vessel recanalization occurs between 21 and 35 days (3–5 weeks) after embolization. It is a nontoxic substance.

Polyvinyl alcohol (PVA) The most commonly used preparation is in the form of PVA particles [Ivalon (Nycomed), Contour (Merck)] which are packed according to size, for example, 150–300, 300–600, 600–1000 and 1000–1500 μm. The particles must be suspended in contrast medium prior to delivery to render the mixture radioopaque. The particles clump readily and can occlude the delivery catheter. Delivery in a suspension of saline and contrast medium in small aliquots and clearing the catheter with 'clean' contrast medium is necessary. If the catheter becomes blocked the gentle passage of a guidewire will usually help to dislodge the PVA plug. Embolization is mechanical (PVA is relatively inert) and vessel occlusion is permanent.

The choice of which particle size to use is based on an estimate of the size of the vessels to be occluded at the centre, or nidus of a lesion. Frequently this has to be educated guesswork. Polyvinyl alcohol particles are not suitable for the primary embolization of large vessel A–V communications of any type. Polyvinyl alcohol mixed in contrast medium with gelatin sponge provides a useful embolic cocktail.

Lyophylized human dura The author does not now use this substance as there have been reports of the development of Jakob–Creutzfeldt disease following its use in neurosurgery (Anon, 1987). Although there have been no such reports from its use as an embolic agent, there are suitable safe alternatives and therefore it is now not recommended.

Mechanical devices

Coils The original embolization coils designed by Wallace *et al.* (1976) have been greatly modified and refined particularly in terms of range of sizes available and ease of delivery.

Coils come preloaded in a cartridge which is delivered into the hub of the catheter. The coil is then pushed into the catheter with a guidewire and delivered until it is extruded from the end. It is important not to use a coil that is too small otherwise it will pass too distally or even through a lesion, whereas a coil that is too large will not form its shape and will elongate within the vessel.

It is advisable to check that the guidewire will pass easily through the catheter without dislodging it before delivering the coil. It is also useful to prime the catheter with saline prior to delivery as this lubricates it and eases passage of the coil.

Ideally, coils are delivered through polyethylene catheters. They can be delivered through polyurethane but there is significant abrasion of the inside of the catheter and, consequently, there is a much greater tendancy for the coil to stick. If a coil is firmly stuck within the catheter shaft the entire assembly can be removed. If however it is partially extruded and stuck it can be more difficult. Withdrawal of the whole assembly runs the risk of wiping off the coil from the end of the catheter and embolizing an adjacent organ or 'losing' the coil in a limb. It is sometimes possible to force the coil out by a forceful injection of saline through a small syringe (increased force per unit area). If it remains impossible to free the coil a long catheter sheath is passed over the entire catheter, the catheter and coil withdrawn into it, and the whole assembly withdrawn.

If a coil is inadvertently 'lost' into another organ or limb the interventional radiologist should attempt removal (see the chapter on Retrieval of Intravascular Foreign Bodies).

Coils have wool or Dacron threads attached to them which act as a nidus for thrombus formation. Vessel occlusion is permanent. It is frequently necessary to produce a nest or string of coils in the lesion or vessel to be occluded in order to provide a meshwork for thrombus formation.

The versatility of coils is increased by the use of the Marsmann wire. This wire has a central movable mandril which is longer than the outer wire. The coil is loaded onto the exposed distal mandril and it can then be delivered exactly to the point desired before being released into the vessel.

Coils are particularly useful in the occlusion of large vessels especially in renal pathology, AVMS (systemic or pulmonary), postbiopsy/traumatic A–V fistulae, varices and spermatic variocoeles.

Balloons (Novak, 1990b) The indication for the use of balloons are the same as for coils, i.e. situations where radioopaque embolic agents of predetermined size and configuration are required, for example, pulmonary AVMs. They are also extensively used in the central nervous system in the management of, for example, caroticocavernous fistulae, AVMs, unclippable aneurysms etc.

Various detachable balloon catheter systems have been developed and are commercially available. Each system has its own peculiarities relating to delivery, inflation and detachment. The reader is advised to consult the literature if balloons are to be used, to read the manufacturer's instructions carefully and ideally to arrange *in vitro* and *in vivo* demonstrations of their use.

Balloons are generally more expensive than coils and in the non-neurological context the indications for their uses are virtually identical. Coils are cheaper and easier to handle and therefore tend to be more widely used.

Liquid embolic agents

All liquid embolic agents are clear and virtually colourless. They are not radioopaque, and are difficult if not impossible to control once injected into the vascular tree. They will be carried into the smallest vessels, desirable in many therapeutic situations but also potentially hazardous if they permeate into vulnerable vascular beds. They must therefore be used with great caution.

Hypertonic dextrose Hypertonic (50%) dextrose solution can be used alone or in combination with other agents as part of a cocktail. It is intensely irritant to the endothelium and causes thrombosis in the vessels into which it is directly injected. Like many other liquid agents dextrose causes significant pain when injected. The patient must be warned of this, a sudden inadvertent movement by the patient could dislodge the catheter.

Absolute ethanol Absolute ethanol causes permanent vessel occlusion and organ or tissue infarction. The agent causes intense vascular spasm and rapid occlusion of small peripheral vessels. Alcohol also causes endothelial damage and activation of the coagulation system. Many authorities advocate the use of occlusive balloon catheters for the delivery of alcohol to prevent reflux into adjacent areas. It must be remembered that when an occlusion balloon is used the natural haemodynamics are changed and that on deflation of the balloon material distal to the balloon may be washed out by the now in-flowing blood.

Alcohol has been used for renal ablation (tumours, hypertension), variceal embolization and varicocoele treatment. Direct percutaneous injection of peripheral AVMs has also been reported (Yakes *et al.*, 1986, 1989).

Isobutyl-2-cyanoacrylate (Bucrylate) This agent is mentioned simply to warn the novice to avoid using it at all costs. It is exceptionally difficult to handle, setting virtually instantaneously on contact with body fluids, is classified as a suspect carcinogen and its use is limited to the management of life-threatening situations only, for example, exsanguination (Novak, 1990b, Allison and Hemingway, 1992).

Chemoembolization

Embolic agents can be mixed with or tagged to chemotherapeutic agents or isotopes to increase local concentration and efficacy and reduce systemic toxicity.

TECHNIQUE

Clinical liaison

As with other interventional procedures close clinical liaison is imperative. This ensures appropriate indications for referral, clinical workup and follow-up. It also ensures that in the event of a complication adequate clinical backup is available.

Examination

The operating radiologist should see and examine the patient prior to the procedure. This allows an adequate explanation of the procedure to be given to the patient and for the radiologist to plan the most appropriate procedure. The examination should include the assessment of the femoral pulses, the patient's general medical state, history of allergy, diabetes etc. or other relative contraindications to the procedure.

Informed consent

It is advisable for the radiologist to obtain consent from the patient whenever possible. An adequate explanation of the procedure, including the significant risks should be given before the patient arrives in the radiology department and before any form of premedication has been given. Clinical house officers cannot be expected to give the necessary information to the patient or to answer any questions the patient may have relating to new and complex interventional procedures.

The author strongly recommends the reader to consult the authorative discussion of this subject by Allison *et al.* (1992).

Surgical cover

A wide variety of potential complications of vascular procedures can occur *(vide infra)* therefore it is imperative that an appropriate surgical team is aware when and where procedures are occurring and are available for consultation and emergency treatment if necessary.

PREMEDICATION AND ANAESTHESIA

The vast majority of embolization procedures are performed using local anaesthesia. Lignocaine, either as 1 or 2% solutions (not exceeding the maximum dose in the adult of 200 mg), is injected subcutaneously and into the perivascular tissues. It is important to encircle the artery/vein with anaesthetic as this ensures good anaesthesia and significantly reduces arterial spasm.

Some adult patients and most children will require general anaesthesia if they cannot remain still and cooperate; the procedure can be painful and surroundings can be very frightening.

Premedication is important to allay anxiety and minimize discomfort. A useful combination is an anxiolytic such as a benzodiazepine and an analgesic such as pethidine. Doses vary according to age and body weight. *However*, it is very important to remember that these drugs singly or in combination can produce cardiorespiratory depression. Patients must be carefully monitored, particularly in the interventional suite which is often dimly lit where changes in patient's colour and vital signs can be difficult to observe. In heavily sedated patients pulse oximetry is advised, and full resuscitation equipment, including drugs to reverse sedation and analgesia must be available at all times (Report of a Joint Working Party of the Royal Colleges of Anaesthetists, 1992), for example, Flumazenil for the reversal of sedative effects of benzodiazepines and Naloxone hydrochloride for the reversal of opioid-induced respiratory depression.

PRELIMINARY ANGIOGRAPHY

Prior to performing embolization it is essential to perform high-quality preliminary angiography to delineate the local vascular anatomy, the specific vascular supply to the lesion under evaluation and to assess the feasibility of the procedure being planned (Allison and Machan, 1992).

Arteriography is usually performed via a percutaneous transfermoral approach. Vascular access is gained using a needle/guidewire/catheter exchange technique as originally described by Seldinger. This will not be described in detail here but the reader is referred to the description given in

Chapter 1. An appropriate catheter *(vide supra)* is introduced and guided under fluoroscopic control to the area of interest. Iodinated radiographic contrast medium is then injected into the vessels and the information recorded either on film or computer as in DSA *(vide supra)*. It is advisable to use nonionic low osmolality contrast medium as embolization procedures, particularly in the case of AVMs, can be lengthy and require a large volume of contrast medium to complete them. It is also important to ensure that the patient is well hydrated.

All vessels supplying a lesion must be demonstrated so that all potential routes of collateral supply are known and coexistent supply to adjacent vulnerable vascular beds is recognized. These studies also form the baseline against which the progress of treatment is evaluated. If a lesion is considered suitable for embolization then it may be possible to carry this out at the same procedure (providing appropriate consent has been obtained). The catheter for embolization is then placed superselectively in the vessels to be occluded. Ideally the catheter should not be wedged, and there should be no reflux back around the catheter on a gentle hand injection of contrast medium. The appropriate embolic agent(s) are then injected under fluoroscopic control and the progress of the procedure monitored by repeated 'mini' angiograms. A final angiogram should always be obtained at the end of the procedure to record the appearances after treatment.

The choice of embolic agent

Embolic agents have been discussed in detail earlier in the chapter. It is worth reiterating the very important principle of embolization that whichever agent is being employed it is vital to deliver it as close to the nidus of the lesion as possible. A proximal delivery of an occlusive agent or device in a vessel which is supplying either a bleeding point, an AVM or a tumour will give good angiographic results but will afford the patient very little if any therapeutic benefit and indeed will possibly preclude further more effective attempts at embolization. The obliteration of the nidus of a lesion significantly reduces its potential to revascularize and hence to recur.

Specific indications

Systemic arteriovenous malformations

(Allison and Hemingway, 1986, 1992)

The treatment of AVMs by arterial embolization has proved to be one of the most successful applications of the technique (Hemingway *et al.*, 1987). Generally speaking, lesions that exhibit a significant element of arteriovenous shunting (fast flow lesions that are pulsatile with a bruit on auscultation) are most amenable to treatment in this way (Gomes *et al.*, 1983) but those that are predominantly capillary or venous in nature (nonpulsatile, slow flow lesions) are not.

Arteriovenous malformations may present in a wide variety of ways including disfigurement, local swelling, pressure symptoms, pain, haemorrhage, venous congestion, distal ischaemia or, in the case of large lesions, cardiac strain and high output cardiac failure (Fig. 3.2). Small lesions can usually be totally excised but surgery on large or deep-seated lesions may be difficult or impossible by virtue of their vascularity, and formerly the only effective treatment for very large AVMs in a limb was amputation. Ligation of feeding vessels is ineffectual and produces only temporary control of such lesions. Partial excision is useless as previously undetected or inaccessible collateral vessels rapidly enlarge and

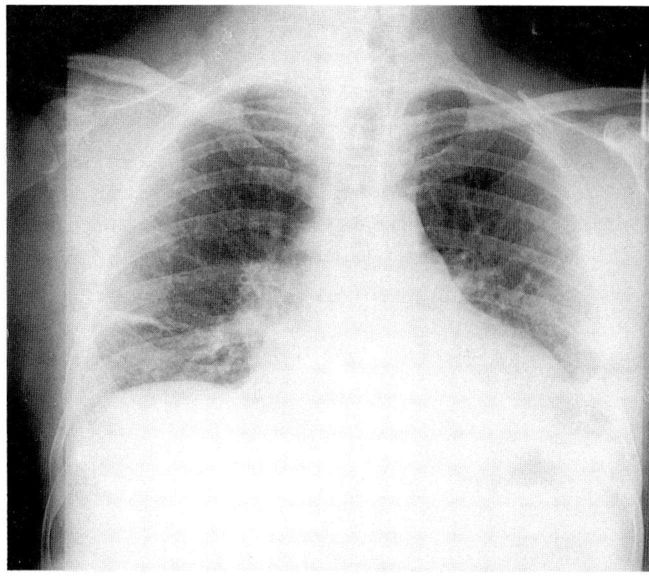

Fig. 3.2
Chest radiograph of a 58-year-old man with cardiomegaly and high output cardiac failure secondary to a massive arteriovenous malformation in his shoulder and right arm.

maintain the blood supply to the abnormality. The introduction of therapeutic arterial embolization has significantly changed the management of these lesions in all areas of the body. Ideally all patients referred for treatment of an AVM are assessed in an outpatient clinic conducted jointly by a radiologist and a vascular surgeon. A detailed clinical history is taken, a history of trauma or previous surgery is very important and physical examination and appropriate investigations including palpation, auscultation, Doppler ultrasound and plain film radiography are performed (Young, 1988). Many patients, particularly those with primarily venous lesions, are advised not to undergo any further investigations or treatment since these lesions (though unsightly) rarely pose a threat to life or limb and the results of embolization are extremely disappointing. Local sclerosant therapy is sometimes helpful in such cases.

Arteriography should be performed in those patients in whom a significant arterial input to the lesion is detected by examination and Doppler ultrasound. The relative risks, hazards and benefits of embolization are explained to the patient (and relatives if appropriate) prior to angiography so that the diagnostic and therapeutic procedure can be performed on the same occasion if appropriate.

PRELIMINARY ARTERIOGRAPHY (Fig. 3.3a–d and 3.4a–b)

High-quality arteriography (with nonionic, low-osmolar contrast media) is crucial in patients with AVMs. It is essential to demonstrate all feeding vessels, all potential sources of collateral supply, the size of any major arteriovenous communications and the draining veins. It is also necessary to establish the extent to which arteries communicate directly with the venous system in the malformation, as opposed to breaking down into a capillary bed as this will significantly influence the size and type(s) of embolic agent used to occlude the lesion.

It must also be established whether there are any communications with adjacent vulnerable vascular beds. In the limbs it is important when injecting selectively into a malformation to determine whether or not contrast medium appears in the distal vessels of supply to the limb; if it does then the potential exists for emboli to pass through the same route and cause distal limb embolization and the procedure may then be contraindicated.

RECONSTRUCTIVE SURGERY

Occasionally angiography will reveal that a lesion is unsuitable for embolization; this is particularly true in the case of lesions affecting the extremities where it may be apparent that effective embolization can only be achieved at the cost of possibly losing one or more digits. Some patients are prepared to accept this risk as their symptoms of pain and ischaemia due to a steal phenomenon are too disabling to tolerate (Fig. 3.5).

In some patients with AVMs previous surgical procedures may have resulted in the ligation of one or more of the major feeding vessels leaving the lesion supplied by numerous tiny collaterals which cannot be successfully catheterized and embolized. It is possible on occasion to reconstruct vascular access for the embolist by restoring flow to a major feeding vessel by a vein graft (Strachan *et al.*, 1986). This may cause temporary enlargement of the lesion but an effective embolization can then be performed through the graft. In other instances it may be valuable to bypass a lesion with a graft; the graft ensures continuity of supply to normal vessels beyond the malformation and, by ligating the native vessel above the lower anastomosis, embolization of the malformation can then be performed without the risk of any distal embolization to normal vessels.

EMBOLIZATION (Flye *et al.*, 1983; Riché and Merland, 1988; Widlus *et al.*, 1988; Roche, 1990)

When a lesion is considered suitable for embolization, feeding vessels are catheterized superselectively and suitable embolic agents are injected into the 'nidus' of the lesion. The embolic material should occlude the lesion completely (Fig. 3.6) but at the same time not pass through into the venous system and cause pulmonary embolism. It is also important not to deliver a large embolus proximally before the nidus of the lesion is occluded, otherwise a situation analogous to, and no more effective than, a surgical ligation will be created and collateral vessels will develop very rapidly distal to the point of proximal occlusion (Figs 3.7 and 3.8). The most useful materials for embolization are Gelfoam particles, PVA, hypertonic dextrose and steel coils. Great caution must be exercised when using liquid agents as these can permeate through the smallest collaterals and, although often the most effective occlusive agents, they are also potentially the most hazardous.

Sometimes it may be useful to devascularize a lesion by embolization prior to definitive surgical resection (Vaughan *et al.*, 1985). This is a useful combined approach in large but localized lesions, for example, on the buttock or in the neck. Embolization is a relatively safe procedure in experienced hands, leaves no scars, is readily repeatable, well tolerated by patients and at no stage precludes surgery if this is thought desirable.

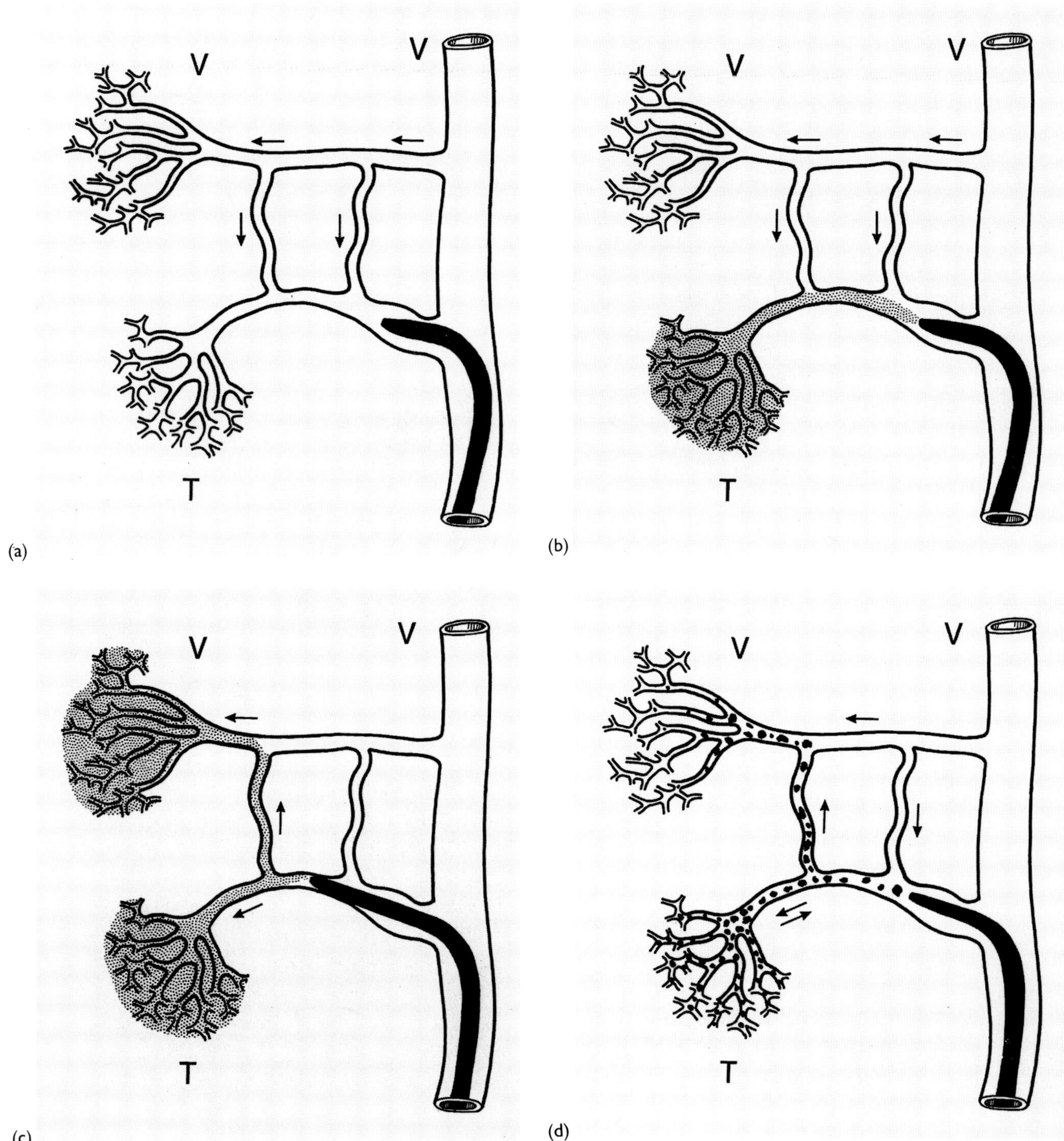

Fig. 3.3
(a) A catheter is placed selectively in a vessel supplying a target (T) lesion which is also supplied by collaterals (↓) from an adjacent vulnerable (V) vascular area. (b) Contrast medium (shaded area) injected from this catheter position may not demonstrate the collaterals as flow is predominantly into the target (T) lesion. (c) A more selective or wedged catheter position may alter the haemodynamics and demonstrate the communications with the vulnerable (V) area. (d) Similarly emboli once injected will occlude the small vessels within the target (T) lesion, change the local haemodynamics and there will then be reversal of flow in the communicating vessels allowing passage of emboli into the vulnerable area. Reproduced with permission from Professor D.J. Allison.

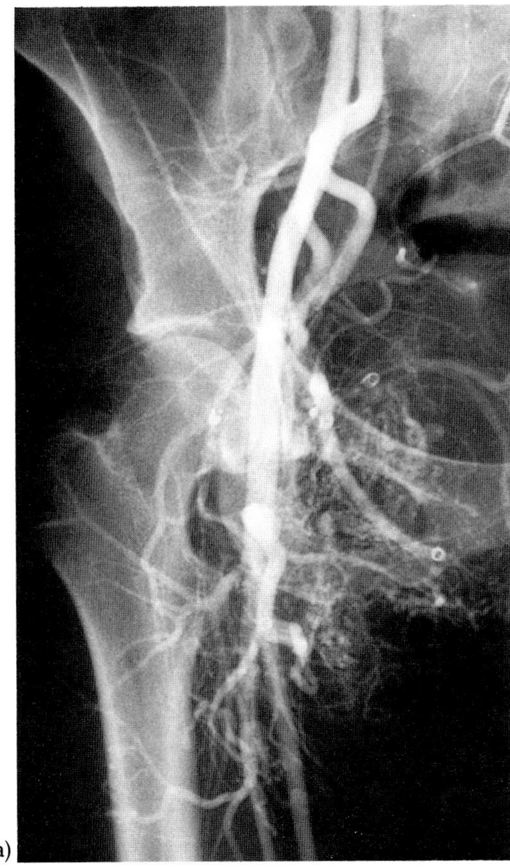
(a)

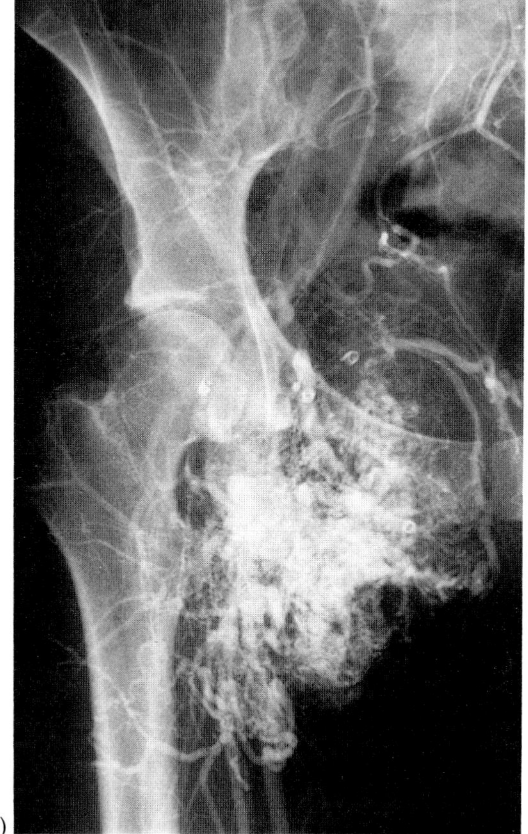
(b)

Fig. 3.4
Arterial (a) and capillary (b) phases of a pelvic arteriogram in a girl with a large gluteal arteriovenous malformation supplied not only by ipsilateral branches of the internal iliac and profunda femoris arteries but also via branches from the contralateral internal iliac.

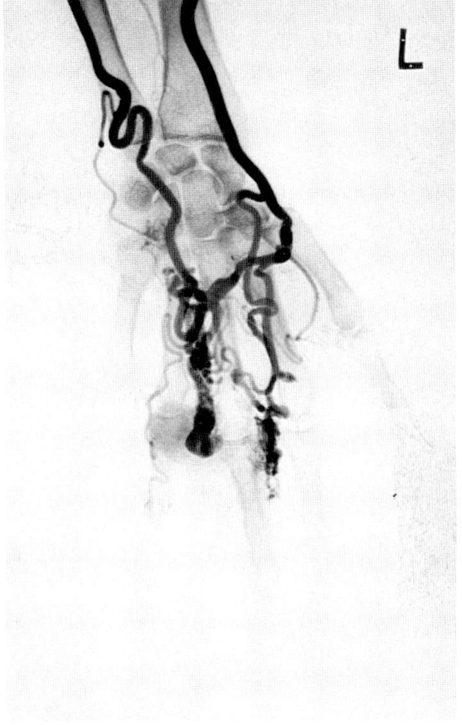

(a)

(b)

Fig. 3.5 (a) Arterial phase. (b) Venous phase
Brachial arteriogram in a woman with a large arteriovenous malformation affecting the palm and third and fourth digits. Effective embolization of all of the small vessels supplying the lesion would inevitably seriously jeopardize the viability of the digits. Reproduced with permission of Professor D.J. Allison.

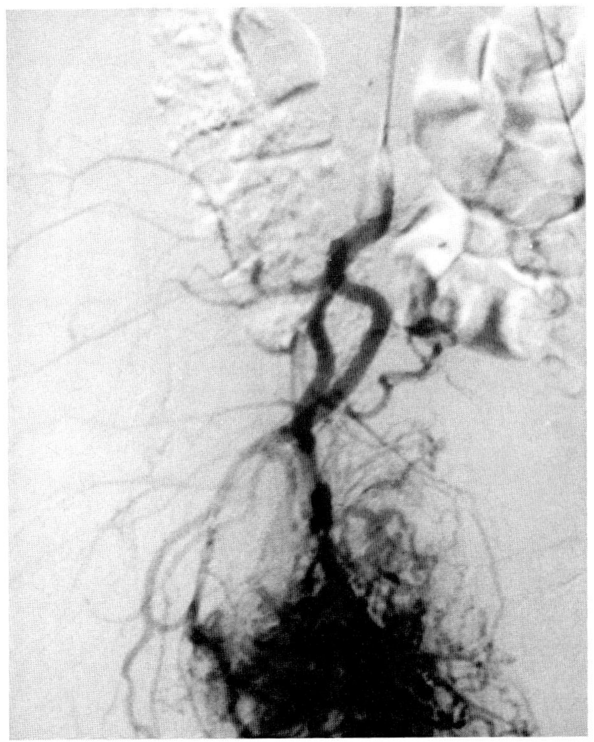

(a)

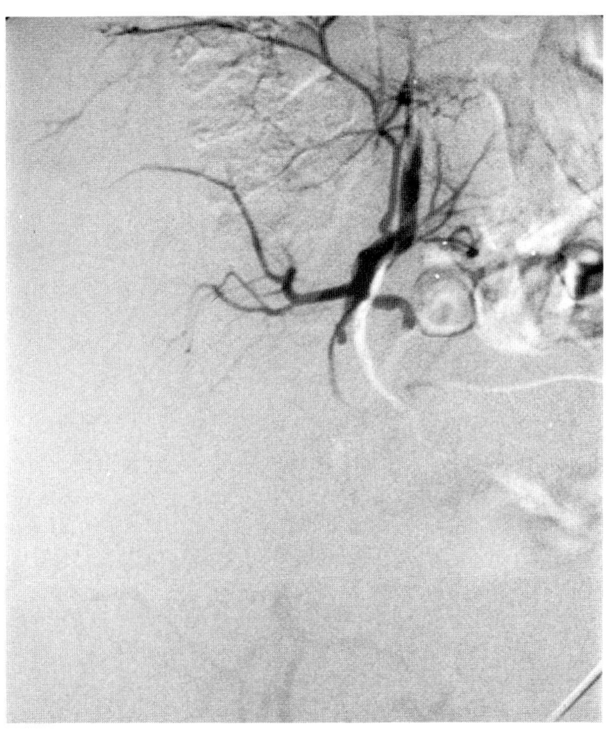

(b)

Fig. 3.6
(a) Pre-embolization and (b) postembolization internal iliac arteriograms in the same patient as Fig. 3.4 after selective catheterization of both of the internal iliac branches demonstrated in (a) The nidus of the lesion was occluded with Ivalon and Spongostan particles. The postembolization study shows flow in normal proximal vessels but complete occlusion of the lesion.

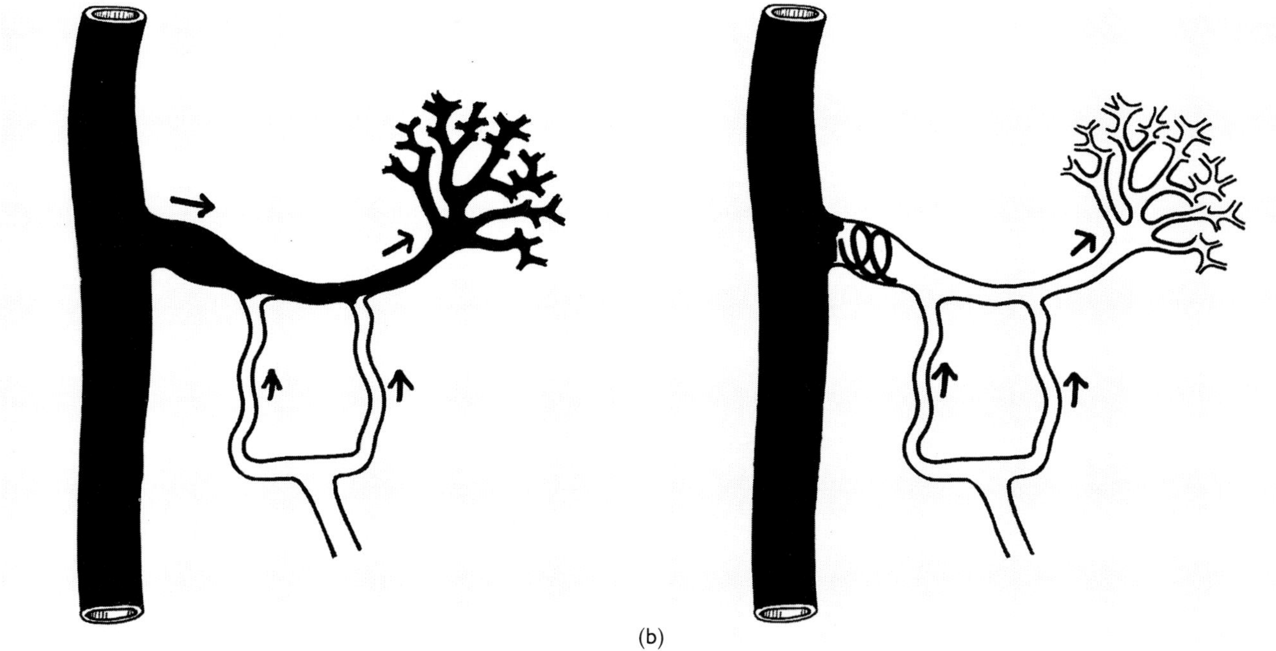

Fig. 3.7
(a) Contrast medium injected into a major vessel demonstrates the main vessel supplying a lesion (←), but not the smaller collaterals (↑).
(b) A coil placed proximally in the main vessels results in a good angiographic result but the lesion continues to be supplied by the collaterals (↑). A situation analagous to a surgical ligature. Reproduced with permission of Professor D.J. Allison.

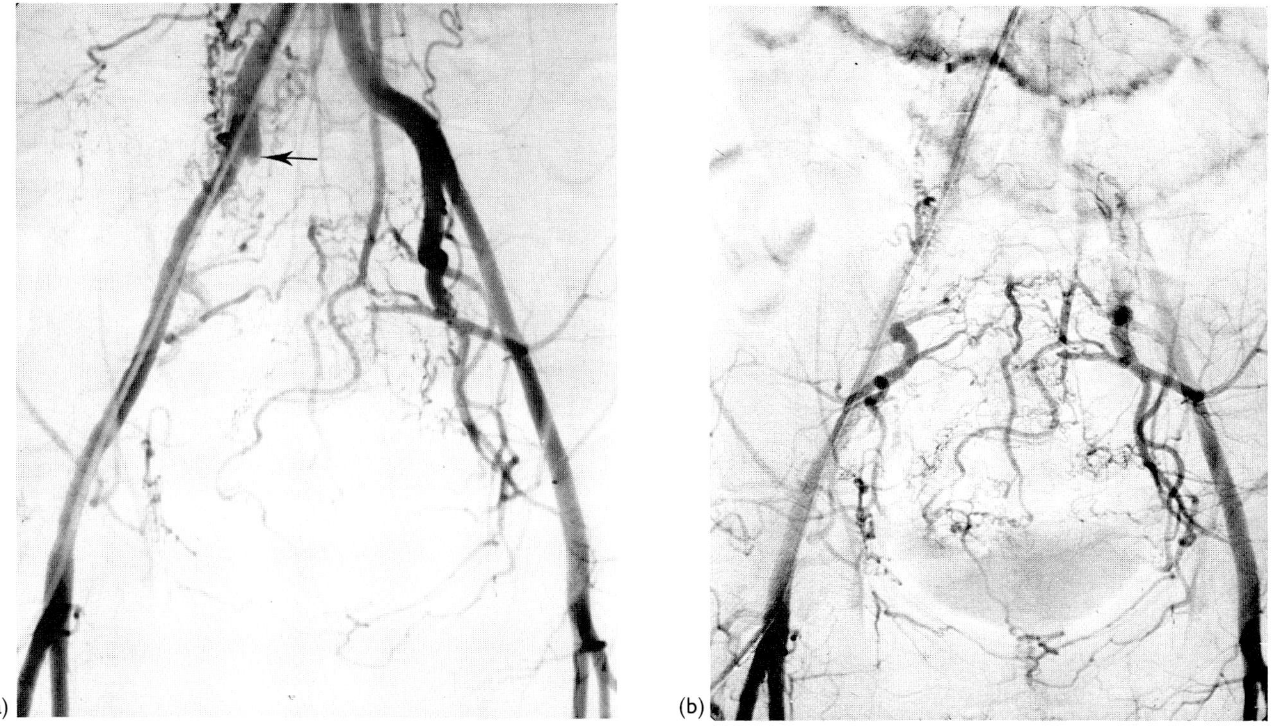

Fig. 3.8
Pelvic arteriogram (a) early and (b) late phases in a woman with a pelvic arteriovenous malformation which had been previously treated surgically by ligation of the right internal iliac artery (RIIA). In the early phase (as) the RIIA stump is seen (arrow), in the later phase the distal RIIA is clearly visible supplied by numerous ipsiateral and contralateral collateral vessels. The ligature has effectively only removed a short segment of main vessel without any significant reduction in vascularity distally. Reproduced with permission of Professor D.J. Allison.

FOLLOW-UP

All patients who have had AVMs of any appreciable size embolized should be reviewed regularly at a combined surgical–radiological clinic, and re-evaluation and repeat embolization(s) performed as necessary.

TRAUMA

Traumatic injury to blood vessels may result from penetrating injury such as gun shots, stab wounds, accidents or blunt injury such as road accidents or sports injuries (horse riding, ski-ing etc.). Iatrogenic injury following surgery, biopsy and interventional procedures must also be remembered (Sclafani *et al.*, 1986).

Angiography and embolization may play a role not only in the immediate assessment of trauma but also in the management of delayed complications.

Acute vascular injury

Arteriography in the acute situation can establish the nature of the vascular damage sustained and delineate its extent. At the same time the study provides an invaluable anatomical vascular roadmap for the surgeon. In some patients profuse haemorrhage can be controlled either preoperatively or definitively by embolization and this can be a life-saving manoeuvre, particularly in patients whose general condition or whose injuries make immediate surgery inappropriate or dangerous. The angiogram is also extremely valuable in excluding major vascular injury and therefore preventing unnecessary exploratory surgery in patients in whom serious diagnostic doubt exists.

Signs which are sought on such a study include extravasation of contrast medium indicating active haemorrhage, absence of filling of a vessel suggesting avulsion from its trunk or severe spasm, marked changes in calibre of a vessel, overt abnormalities such as aneurysms or arteriovenous communications, or disruption of the continuity of a vessel. It is important to remember venous injury as well as arterial damage. The only angiographic evidence of an arteriovenous fistula may be early venous filling and this sign should be carefully sought. If active haemorrhage is demonstrated this may be controlled by therapeutic embolization. This technique has proved to be of value in the pelvis when bleeding is from small vessels which may be very difficult or impossible to identify and control surgically (Hansen and Kadir, 1991).

If angiography and embolization, in the management of trauma, are to be of value to the surgeon, it is important that they are available as an emergency service on a 24 hour basis.

Delayed complications of vascular injury

Vascular damage may not manifest itself for days, weeks or even months following trauma. In these situations the most common occurrence is the development of a false aneurysm at the site of injury with or without an associated arteriovenous fistula, an event that can happen following either penetrating injuries or blunt trauma. The aneurysm may reveal itself as a pulsatile mass (Fig. 3.9), it may rupture causing life-threatening haemorrhage or, in the case of an arteriovenous fistula (Fig. 3.9) symptoms may appear referable either to elevated venous pressure, cardiac failure or ischaemia distal to the site of shunting.

Arterial lesions presenting as delayed complications of traumatic injury are frequently amenable to embolization. Ideally, the catheter should be advanced to the neck of the aneurysm or fistula and the lesion embolized as selectively as possible. Embolization restricted to more proximal feeding arteries carries the significant risk that the lesion will continue to receive blood from collateral feeding vessels. Angiography sometimes reveals a lesion that, while theoretically suitable for embolization, would be more appropriately treated by surgical ligation or excision, for example, a communication between the common femoral artery and vein where the fistula may be wide necked and short making it difficult to lodge an embolus within it safely without jeopardizing either artery or vein (Altin *et al.*, 1989). Chronic arteriovenous fistulae in the extremities may be virtually indistinguishable from congenital lesions in appearance having gained numerous vessels of supply over time (Lawdahl *et al.*, 1989).

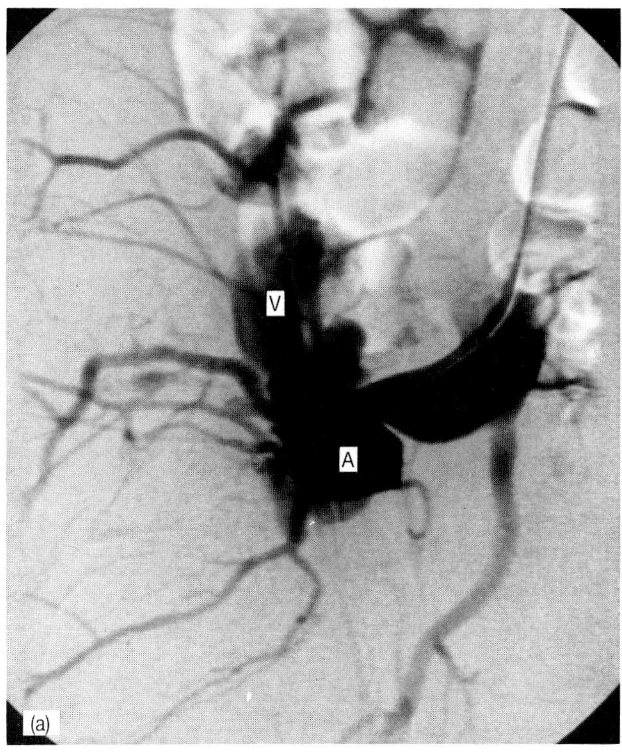

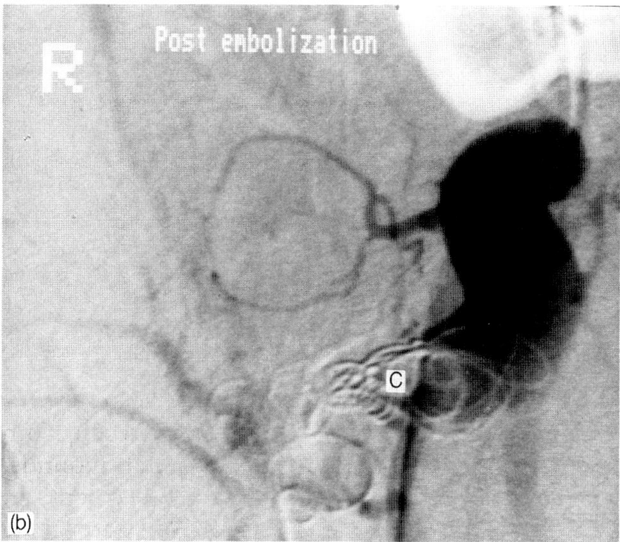

Fig. 3.9
(a) Selective right internal iliac arteriogram in a man who had suffered blunt pelvic trauma. The main vessel is markedly enlarged with aneurysmal dilation distally (A) and rapid filling of a large vein via a fistula (V). (b) The aneurysm and fistula were effectively occluded by the placement of a number of coils (C), leaving normal proximal vessels patent.

Tumours

Embolization has gained wide acceptance in the management of certain forms of tumour. The technique can be used, definitively (benign lesions only), preoperatively or palliatively. Embolization has been used to treat tumours in all areas of the body. The principal indications for its use are haemorrhage, pain, tumour bulk causing local pressure symptoms and excessive hormone secretion (Carrasco, 1991). The basic principles of the technique are the same as in AVMs and trauma. It is puzzling why tumour embolization has not gained a more significant place in the UK compared to Europe and the USA. Certainly vascular radiologists working in centres with large oncology and radiotherapy departments should be prepared to offer and develop this service.

Definitive embolization

Some benign tumours can be treated successfully by embolization alone. The technique has been reported to be useful in the management of tumours of vascular origin (haemangiomas, angiomyolipomas) in a variety of sites including bone. A variety of benign bone lesions have been treated successfully by embolization, and in cases of large lytic lesions the procedure may help to obviate the risk of pathological fracture. There are a number of reports of successful management of aneurysmal bone cysts by embolization (Chuang *et al.*, 1982; Suby-Long *et al.*, 1988).

Preoperative embolization

Embolization has been employed preoperatively in bone tumours (prior to limb preservation surgery). In this situation the embolization may be performed purely to reduce vascularity or may be combined with chemoembolization to provide acute chemotherapy directly to the lesion prior to surgery (Stine and Falletta, 1991).

Immediate preoperative embolization may also be valuable in the devascularization of soft tissue tumours such as haemangiopericytomas.

Ideally surgery is carried out within 24 hours of the embolization. A delay of a few days can result in a more complex surgical procedure related to the significant oedema which can occur surrounding an embolized lesion.

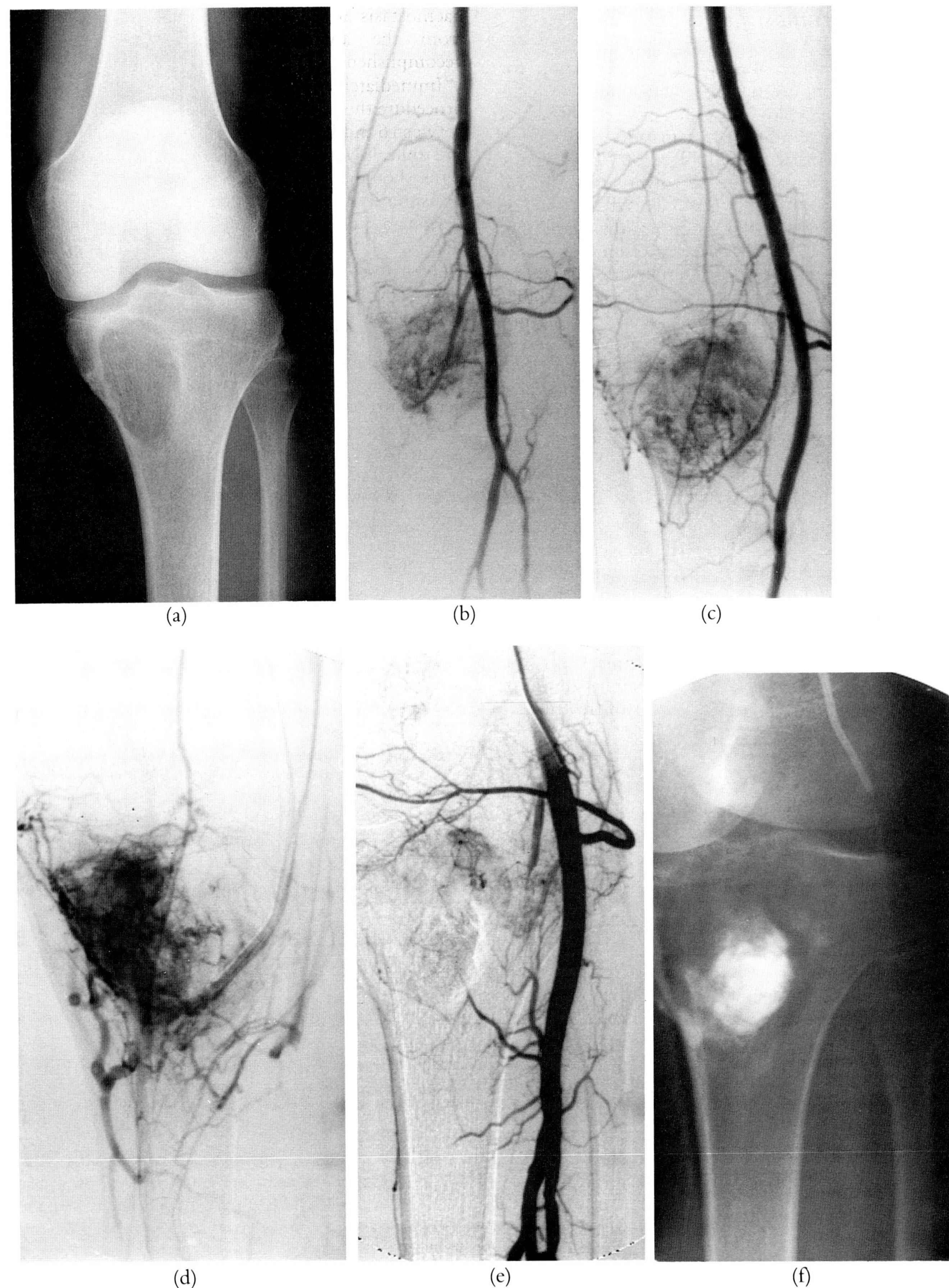

(a) (b) (c)

(d) (e) (f)

Palliative embolization

This is by far the most common indication for embolization in tumour management. The technique can be employed as the primary mode of treatment in inoperable neoplasms and some studies have suggested that the embolization of metastatic deposits, for example, in the liver, may extend survival time in patients with advanced disease. The main purpose of embolization in this situation, however, is not to increase survival times but rather to improve the quality of life by relieving distressing symptoms such as haemorrhage, pain, tumour bulk, venous obstruction, endocrine effects etc. A significant amount of work has been performed and is continuing in investigating the value of labelling embolic agents with cytotoxic agents, monoclonal antibodies or radioactive isotopes to improve the efficacy of the procedure.

Bone neoplasms, both primary and metastatic (Feldman *et al.*, 1975) also appear to respond well to treatment by embolization. Pain from primary and metastatic lesions may be dramatically relieved and bone regrowth may be observed. Lesions that have been successfully managed in this way include giant cell tumours (Chuang *et al.*, 1982) and metastases from renal cell carcinoma and breast (Fig. 3.10).

Embolization should be considered in the management of any hypervascular primary or metastatic neoplasm that is causing intractable symptoms unresponsive to conventional means of treatment.

Aftercare and follow-up

A significant number of patients require two or more procedures, thus it is important to prevent haematoma formation at the puncture site as scarring makes subsequent procedures more difficult. Haematoma is also associated with false aneurysm formation. The operator or other radiologist must be responsible for groin compression and haemostasis and the patient should not be moved from the angio suite until this has been accomplished.

Immediately following an embolization procedure the patient should be kept on bed rest for between 6 and 12 hours to allow the arterial puncture site to heal over. During this time the patient's vital signs should be monitored including pulse, blood pressure, hydration and the vascular puncture site. Peripheral pulses should also be monitored both in the limb of the arterial puncture and also the limb that has the lesion which has been treated. Vascular spasm, haematoma formation at the puncture site and significant swelling of the embolized lesion secondary to thrombosis and oedema can all occur and should be reported to the radiologist and the responsible clinician immediately as further intervention or treatment may sometimes be necessary to prevent a serious complication.

The patient's temperature must also be carefully monitored and regular checks on the white blood cell count (WBC) made as infection in the embolized tissue leading to abscess formation and septicaemia must be detected and treated rapidly. The use of prophylactic antibiotics is controversial, but advisable if bone lesions are embolized. Any patient in whom a significant volume of tissue has been rendered necrotic will develop a fever and elevated WBC as part of the postembolization syndrome *(vide infra)*, but a sustained fever, or a spiking pyrexia should raise the suspicion of infection and be investigated accordingly with serial blood cultures. Estimation of the serum C-reative protein (CRP) is useful as an indicator of infection (Hind *et al.*, 1985). The 'normal' response to embolization is represented by a rise in CRP, reaching a maximum at 5 days and returning to normal over the next 10 days. A continual rise or a failure to fall indicates infection and should be investigated and managed accordingly.

Fig. 3.10
(a) Plain AP radiograph of the knee in a male patient with a painful lytic metastasis from a primary renal carcinoma. (b) AP and (c) lateral superficial femoral arteriograms showing that the lesion is hypervascular and supplied predominantly by inferior geniculate vessels. (d) This vessel was superselectively catheterized and (e) occluded with a mixture of Ivalon, Spongostan and hypertonic dextrose mixed with contrast medium. (f) A plain radiograph after embolization shows stagnant contrast medium stained embolic material and blood within the lesion.

Complications

(Miller and Mineau, 1983; Hemingway and Allison, 1988; Novak, 1990a; Hemingway, 1991)

General complications include those related to the angiographic procedure (Hessel *et al.*, 1981), for example, dissection, haematoma, vascular spasm, contrast medium reaction, and those related to any embolization procedure, for example, pain, fever, elevated WBC (the postembolization syndrome, PES, a response to tissue necrosis) infection, distal embolization of nontarget tissues etc., reaction to embolic agent (rare), unintentional tissue necrosis, renal failure (dehydration and massive tissue necrosis) and death (Table 3.3).

Any embolization procedure, no matter how carefully or expertly performed is associated with an element of risk to the patient. This risk may be very difficult to quantify and is dependent on a large number of factors (Tables 3.4 and 3.5).

The postembolization syndrome

The postembolization syndrome (PES) is thought to represent the body's natural response to the presence of necrotic tissue. Characterized by fever, discomfort, leucocytosis and malaise for 3 to 5 days the severity of the PES appears to be related to the volume of tissue embolized, as well as the nature of the underlying disease process. For example, a more severe reaction would be expected following embolization of a large tumour than following the occlusion of a single bleeding point. A sustained fever raises the possibility of infection. Estimation of CRP is useful in this situation *(vide supra)*.

Specific complications

IMMEDIATE

Pain
Pain may be related to the embolic agent (e.g. alcohol or dextrose), the site, size and nature of the lesion being treated.

Inadvertent embolization of normal structures
This can arise following the incorrect choice of embolic agent, inappropriate catheter position or failure to recognize vascular communications with vulnerable structures.

Pulmonary embolization
Particulate emboli which pass through arteriovenous communications within a lesion will become lodged within the safety net of the pulmonary circulation. This is clearly undesirable, and with careful choice of embolic agent symptomatic iatrogenic pulmonary embolization is fortunately uncommon. When it does occur conservative treatment is usually adequate unless the patient is haemodynamically compromised.

General
Usual hazards of selective arteriography

Specific
Immediate
Pain
Inadvertent embolization normal structures
Pulmonary embolization
Reaction to embolic agent
Systemic embolization (PAVM)

Delayed
Pain
Fever
Swelling
Gas in embolized tissue
Infection
Unwanted tissue necrosis
Parasthesia
Retrograde thrombosis
Release of toxic substances
Renal failure
Death

Table 3.3 *Complications of embolization*

Diagnosis
Size of lesion
Site of lesion
Duration of symptoms
Associated disorders, e.g. diabetes etc.
Age
Complexity of vascular supply
Experience of operator
Choice of embolic agent
Quality of clinical care

Table 3.4 *Factors which may affect incidence of complications*

Proper indications for procedure
Experience
Meticulous technique
Aseptic technique
High-quality fluoroscopy and imaging
Nonionic, low osmolality contrast media
Correct choice of embolic agent
Prophylactic antibiotics when applicable
Clinical liaison

Table 3.5 *Avoidance and prevention of complications*

Reaction to embolic agent
Idiosyncratic reaction can occur to any agent but is fortunately rare.

DELAYED

Pain and fever
This is common as part of the PES. Occasionally the pain is severe enough to require a potent analgesia. A sustained, high or fluctuating fever raises the possibility of infection and should be appropriately investigated.

Swelling
Embolization is occasionally followed by transient and marked local swelling, and oedema is of particular importance if it threatens to compromise vital structures (e.g. airways etc.).

Gas
Gas usually only occurs following the embolization of solid organs, gas in embolized tissue is a benign reflection of tissue necrosis. The development of a fluid level, fluctuation and fever suggest secondary infection.

Infection
Infection represents one of the most feared complications of embolization. A high index of suspicion and early investigation and treatment are essential. Meticulous aseptic technique and the use of prophylactic antibiotics where appropriate (when embolizing solid organs and bone), help prevent most infections although sepsis occasionally occurs despite all precautions.

Unwanted tissue necrosis
Embolization is designed to produce 'controlled' tissue necrosis. Occasionally a larger than desired area is infarcted, or adjacent or distant sites are inadvertently embolized. In the limbs this could lead to significant ischaemia and subsequent amputation. All embolic agents can be associated with excessive necrosis but the liquid agents, by virtue of their ability to permeate the smallest vessels, are most likely to be implicated.

Parasthesia
Parasthesia most commonly occurs with the liquid embolic agents and can occur transiently following the use of fine particulate emboli. For example, liquid agents used in the internal iliac territory related to the sciatic nerve may cause irreversible parasthesia.

Retrograde thrombosis
A vessel once embolized has stagnant blood and thrombus within it. Thrombus may continue to be formed after the procedure has finished and may propagate retrogradely. This is of importance if by extension the thrombus either occludes a vital vessel or fragments and embolizes distant structures.

Release of toxic or humoral substances
This is usually only of relevance if very large volumes of tissue or endocrine tumours are embolized.

Renal failure
Renal failure is precipitated by dehydration, hypotension, contrast medium toxicity and massive tissue necrosis. The risk of renal failure can be minimized by adequate hydration and haemodynamic control.

Death
Fortunately death is extremely rare as a direct consequence of embolization. Death tends to occur in those patients who are very sick at the time of the procedure, for example, in those with large tumours or with massive haemorrhage.

REFERENCES

ALLISON DJ, HEMINGWAY AP (1986) Arteriovenous malformations. In: *Diagnostic Techniques and Assessment Procedures in Vascular Surgery*, pp. 403–20. Edited by Greenhalgh RM. New York: Grune and Stratton.

ALLISON DJ, HEMINGWAY AP (1992) Embolization in the management of vascular anomalies, trauma and tumours. In: *Surgical Management of Vascular Disease*. Edited by Bell PRF, Jamieson CW, Vaughan Ruckley C. W.B. Saunders Co., London.

ALLISON DJ, MACHAN LS (1992) Arteriography. In: *Diagnostic Radiology: An Anglo-American Textbook of Imaging*, 2nd edition. Edited by Grainger RG, Allison DJ. Churchill Livingstone, Edinburgh.

ALLISON DJ, WALLACE S, MACHAN LS (1992) Interventional radiology. In: *Diagnostic Radiology. An Anglo-American Textbook of Imaging*, pp. 2329–88. Edited by Grainer RG, Allison DJ. Churchill Livingstone, Edinburgh.

ALTIN RS, FLICKER S, NAIDECH HJ (1989) Pseudoaneurysm and arteriovenous fistual after femoral artery catheterization: association with low femoral punctures. *American Journal of Roentgenology* **152**: 629–31.

ANON (1987) Creutzfeldt–Jakob disease in a patient receiving a cadavaric dura mater graft. *Journal of the American Medical Association* **258**: 309.

BERENSTEIN A, KRICHETT JJ (1979) Catheter and material selection for transarterial embolization: technical considerations. *Radiology* **132**: 631.

CARRASCO CH, CHARNSANGAVEJ C, RICHLI WR AND WALLACE S (1990) Angiographic management of malignant tumours in the throat, abdomen and bones. In

Interventional Radiology. Edited by Dondelinger R, Rossi P, Kurdziel JC and Wallace S. New York: Thieme.

CHUANG VP (1990) Angiographic embolization techniques. In: *Interventional Radiology*, pp. 284–94. Edited by Dondelinger R, Rossi P, Kurdziel J-C, Wallace S. Thieme Medical Publishers, New York.

CHUANG VP, SOO SC, WALLACE S, BENJAMIN IS (1982) Arterial occlusion: management of giant cell tumour and aneurysmal bone cysts. *American Journal of Radiology* **136:** 1127.

FELDMAN F, CASARELLA WJ, DICK HM, HOLLANDER BA (1975) Selective intra-arterial embolization of bone tumours: A useful adjunct in the management of selected lesions: *American Journal of Radiology* **123:** 130.

FLYE MW, JORDAN BP, SCHWARTZ MZ (1983) Management of congenital arteriovenous malformations. *Surgery* **94:** 740.

GIANTURCO C, ANDERSON JH, WALLACE S (1975) Mechanical devices for arterial occlusion. *American Journal of Radiology* **124:** 428.

GOMES AS, BUSUTTIL RW, BAKER JD, OPPENHEIM W, MACHLEDER HI, MOORE WS (1983) Congenital arteriovenous malformations and the role of transcatheter arterial embolization. *Archives of Surgery* **118:** 817.

HANSEN ME, KADIR S (1991) Embolotherapy for bleeding from pelvic trauma, tumors and operation sites. In: *Current Practice of Interventional Radiology*. Edited by Kadir S. BC Decker Inc., Philadelphia.

HEMINGWAY AP (1986) Embolization materials. *Radiology Now* Volume 3, pp. 63–6. Franklin Scientific Projects, London.

HEMINGWAY AP (1991) Complications of embolotherapy. In: *Current Practice in Interventional Radiology*. Edited by Kadir D. BC Decker Inc., Philadelphia.

HEMINGWAY AP, ALLISON DJ (1988) Complications of embolization: Analysis of 410 procedures. *Radiology* **166:** 669.

HEMINGWAY AP, SMITH EJ, ALLISON DJ (1987) Embolization in the management of systemic arteriovenous malformations: An analysis of 137 procedures. In: *Procedures of the European Contress of Radiology*, p. 415. Edited by Silvestre ME, Abecosis F, Viega-Pires JA. Elsevier Science Publishers, Amsterdam.

HESSEL SJ, ADAMS DF, ABRAMS HL (1981) Complications of angiography. *Radiology* **138:** 273–81.

HIND CRK, THOMAS AMK, PEPYS MB, ALLISON DJ (1985) Serum C-reactive protein response to therapeutic embolization: possible role in management. *Clinical Radiology* **36:** 179.

LAWDAHL RB, ROUGH WD, VITEK JJ, MCDOWELL HA, GROSS GM, KELLER FS (1989) Chronic arteriovenous fistulas masquerading as arterio-venous malformations: diagnostic considerations and therapeutic implications. *Radiology* **170:** 1011–15.

MILLER FJ, MINEAU DE (1983) Transcatheter arterial embolization: major complications and their prevention. *Cardiovascular Interventional Radiology* **6:** 141.

NOVAK D (1990a) Complications of arterial embolization. In: *Interventional Radiology*, pp. 314–24. Edited by Dondelinger RF, Rossi P, Kurdziel JC, Wallace S. Thieme Medical Publishers, New York.

NOVAK D (1990b) Embolization materials. In: *Interventional Radiology*, pp. 295–313. Edited by Dondelinger R, Rossi P, Kurdziel J-C, Wallace S. Thieme Medical Publishers, New York.

REPORT OF A JOINT WORKING PARTY OF THE ROYAL COLLEGES OF ANAESTHETISTS AND RADIOLOGISTS (1992) *Sedation and Anaesthesia in Radiology*.

RICHÉ M-C, MERLAND J-J (1988) Embolization of vascular malformations. In: *Vascular Birthmarks*, pp. 436–53. Edited by Mulliken JB, Young AE. W.B. Saunders Co., Philadelphia.

ROCHE A (1990) Peripheral arterio-venous malformations. In: *Interventional Radiology*, pp. 518–28. Edited by Dondelinger RF, Rossi P, Kurdziel JC, Wallace S. Thieme Medical Publishers, New York.

SCLAFANI SJ, COOPER R, SHAFTAN GW *et al.* (1986) Arterial trauma: daignostic and therapeutic angiography. *Radiology* **161:** 165.

STINE KC, FALLETTA JM (1991) Intra-arterial chemotherapy of osteosarcoma and soft-tissue sarcomas. In: *Current Practice of Interventional Radiology*. Edited by Kadir S. BC Decker, Philadelphia.

STRACHAN J, HEMINGWAY AP, MANSFIELD AO, ALLISON DJ (1986) Embolization of an arteriovenous malformation following surgical reconstruction of a previously ligated lingual artery. *Journal of Interventional Radiology* **1:** 79.

SUBY-LONG T, BOS GD, ROSCH J (1988) Biopsy proven eradication of an aneurysmal bone cyst treated by super-selective embolization. *Cardiovascular Interventional Radiology* **11:** 292.

VAUGHAN M, HENNESSY OF, JAMIESON C, HEMINGWAY AP, ALLISON DJ (1985) The pre-operative embolization of vascular malformations. *British Journal of Radiology* **58:** 717.

WALLACE S, GIANTURCO C, ANDERSON JH, GOLDSTEIN HM, DAVIS JL, BREE RL (1976) Therapeutic vascular occlusion utilizing steel coil technique: clinical applications. *American Journal of Radiology* **127:** 381.

WIDLUS DM, MURRAY RR, WHITE RI, OSTRMAN FA, SCHRIEBER ER, SATRE RW, MITCHELL SE, KAUFMAN SL, WILLIAMS GM, WEILAND AJ (1988) Congenital arteriovenous malformations: tailored embolotherapy. *Radiology* **169:** 511–16.

YAKES WF, HAAS DK, PARKER SH, GIBSON MD, HOPPER KD, MULLIGAN JS, PEVSNER PH, JOHNS JC, CARTER TE (1989) Symptomatic vascular malformations: ethanol embolotherapy. *Radiology* **170:** 1059–66.

YAKES WF, PEVSNER R, REED M *et al.* (1986) Serial embolizations of an extremity arteriovenous malformation with alcohol via direct percutaneous puncture. *American Journal of Radiology* **146:** 1038.

YOUNG AE (1988). Clinical assessment of vascular malformations. In: *Vascular Birthmarks*, pp. 114–27. Edited by Mulliken JB, Young AE. W.B. Saunders Co., Philadelphia.

CHAPTER 4

Retrieval of intravascular foreign bodies

Anna-Maria Belli and Anne P. Hemingway

Introduction 82

Types of foreign bodies and sites 82

Complications of intravascular foreign bodies and indications for retrieval 82

Contraindications to retrieval 83

Patient preparation 84

Equipment and retrieval technique 84

Complications 89

Postprocedural care 91

References 91

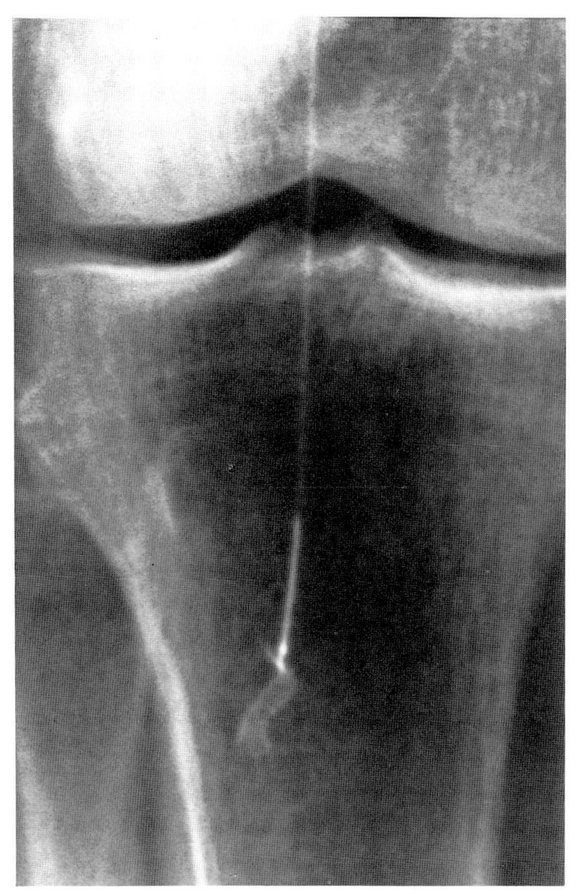

Introduction

Indwelling intravascular catheters were first described by Meyers and nine years later Turner and Sommers reported embolization of an intravenous polyethylene catheter in humans (Meyers, 1945; Turner and Sommers, 1954). Retrieval of intravascular foreign bodies was for many years surgical. Thomas *et al.* (1964) reported the first nonsurgical retrieval and successful retrieval has been achieved over the last two decades by percutaneous fluoroscopically guided techniques. The modern interventional radiologist needs to be able to retrieve items 'lost' in either the venous or arterial system by medical, surgical or radiological colleagues.

Types of foreign bodies and sites

Central venous catheters, pacemakers and monitoring lines introduced by physicians, anaesthetists and cardiologists etc. are indispensable aids to modern patient management. Interventional radiologists introduce a further range of foreign bodies into the arterial and venous system (Table 4.1). Occasionally fragments of (or entire) guidewires, catheters, balloons or embolization coils become detached or lost within the vascular system (Chuang, 1979; Vujic *et al.*, 1986), most commonly during their introduction or removal. Vascular embolization in association with bullet injury has occasionally been reported (Hiebert and Gregory, 1974).

Intravascular foreign bodies (FB) migrate with the flow of blood, so if placed in the venous system they migrate centrally towards the right side of the heart (Fig. 4.1) where they may lodge if made of stiff material or, if softer, they can migrate further into the pulmonary artery and its branches (Uflacker *et al.*, 1986). The final location depends, therefore, on type, size and length of foreign body, its stiffness, site of entry and patient position at time of migration. Arterial foreign bodies tend to be carried peripherally, becoming lodged in a vessel of similar size to the foreign body or at a vessel bifurcation.

Types of intravascular foreign body
Central venous lines
Catheter fragments (arterial or venous)
Metal guidewire fragments
Pacing wires
Ventriculovenous catheters/shunts
Needle fragments
Angiographic dilators
Dilation balloon fragments
Embolization coils/balloons

Table 4.1 *Types of intravascular foreign body*

Complications of intravascular foreign bodies and indications for retrieval

Fisher and Ferreyro (1978) reviewed the literature and found an overall risk of complications of 71% with a 38% risk of death if the intravascular foreign body was not retrieved. Complications of intravascular foreign bodies are shown in Table 4.2 (Doering *et al.*, 1967; Rubinstein *et al.*, 1982). A complication may arise immediately or be delayed for weeks, months or years. Delayed complications are not necessarily less severe (e.g. endocarditis; thrombosis and pulmonary infarction). This coupled with the reported high incidence of complications makes expeditious removal of the retained foreign body mandatory.

Complications of intravascular foreign bodies
Myocardial perforation
Sepsis
Pulmonary embolism/infarction
Cardiac arrhythmias
Cardiac arrest
Endocarditis
Cardiac valve perforation
Pulmonary artery aneurysm (mycotic)
Death

Table 4.2 *Complications of intravascular foreign bodies*

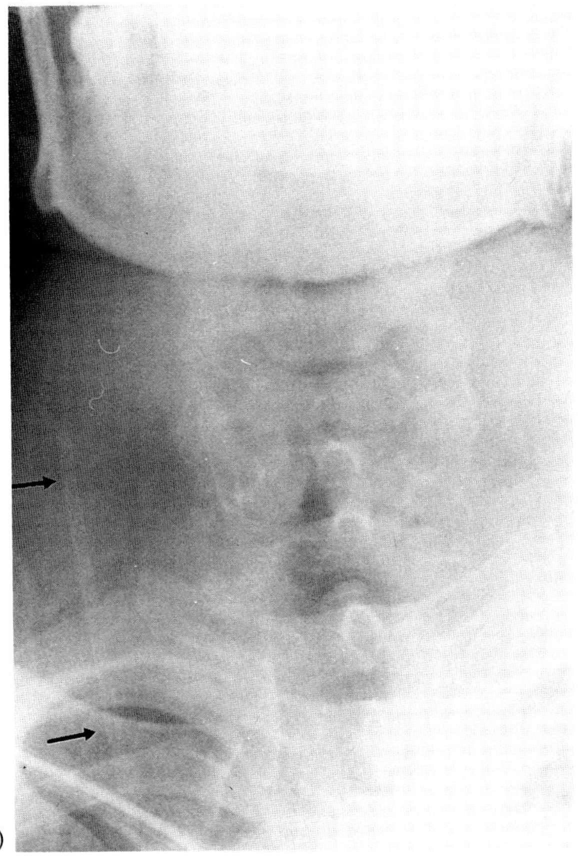

(a)

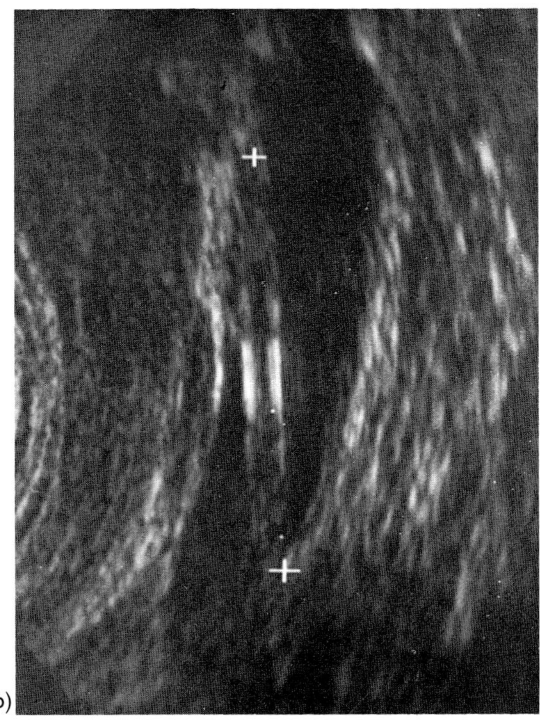

(b)

Fig. 4.1 Broken fragment of vessel dilator in the right internal jugular vein migrating centrally.
Foreign body in the internal jugular vein seen on (a) plain radiography, (b) ultrasound. (c) The foreign body has migrated to the superior vena cava.

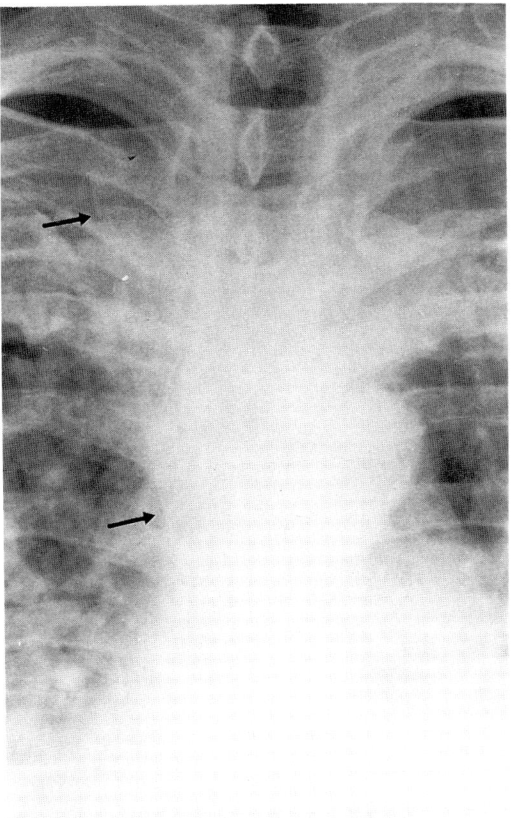

(c)

CONTRAINDICATIONS TO RETRIEVAL

There are no absolute contraindications to attempted removal of retained intravascular foreign bodies. Relative contraindications include a bleeding diathesis, excessive myocardial irritability, distal embolization of a small fragment into the peripheral pulmonary vessels (successful retrieval is unlikely), firm adherence of a fragment to the vessel or cardiac wall (gentle traction NOT sharp pulling is advised), or a large thrombus in the vicinity of the foreign body.

The risks of leaving a foreign body *in situ*, or surgical removal (e.g. thoracotomy), greatly outweigh the risks associated with attempted percutaneous retrieval.

Patient preparation

The cardiovascular and respiratory system should be stabilized before the procedure.

Consent for the procedure should be obtained from the patient if possible, but if the patient is unable to give consent, this should be obtained from the patient's immediate family.

The patient should be well hydrated and on clear fluids only for 6 hours before the procedure to avoid complications should it prove necessary for a general anaesthetic to be administered. However, the procedure should not be delayed if the fragment is continuing to migrate or if there is ongoing cardiac irritability.

Premedication is advised to allay patient anxiety. The patient should be attached to a cardiac monitor to detect cardiac arrhythmias during the procedure. Before commencing the procedure a radiograph may be taken to identify the current position of the foreign body if this is not easily visible on fluoroscopy, or a diagnostic angiogram is performed to locate the exact position of the foreign body. Echocardiography can also be very useful in the detection of an intracardiac foreign body (Davies *et als.*, 1981).

The retrieval procedure should be carried out under strict aseptic technique. In immunosuppressed patients, postoperative patients and when the retrieval procedure has been prolonged, it is advisable to consider antibiotic cover for 2 to 3 days after the procedure.

Equipment and retrieval technique (Table 4.3)

Retrieval is performed under local anaesthesia in most cases and using fluoroscopic control (biplane fluoroscopy is especially useful). The usual approach is via a percutaneous femoral vein puncture for venous foreign bodies. A catheter sheath introducer set is used *(vide infra)*.

Occasionally a jugular approach may be employed for fragments lodged in the superior vena cava (SVC) or subclavian veins to avoid traversing the right side of the heart.

Arterial foreign bodies are usually retrieved from a femoral artery approach.

A radiograph is obtained to confirm position of the foreign body (Fig. 4.1), although sometimes it is more easily recognized on fluoroscopy or ultrasound, particularly if it is not very radio-opaque.

Catheter introducer sheath set
Pigtail/preshaped catheters
Loop snare
Retrieval forceps
Dormia basket
Balloon catheters
Steerable/deflecting/preshaped guidewires

Table 4.3 *Equipment*

Catheter introducer sheath set

This is an essential part of the equipment for foreign body retrieval. The foreign body, when snared or grasped by whichever method has been used, would damage the vein or artery by tearing the vessel if it was simply withdrawn without protecting the puncture site. So the foreign body must be withdrawn into a sheath and the sheath containing the foreign body is withdrawn as a unit. As large a sheath as can safely be inserted is used, so that it can house the foreign body. In the case of a knotted catheter, the knot must be withdrawn into the sheath. If the foreign body is composed of soft, malleable material this is not difficult. If the foreign body is made of harder material, for example, a segment of vessel dilator, this has to be guided carefully into the sheath. It may be helpful to cut the tip of the sheath obliquely with a blade, so that there is a large aperture into which the foreign body can be withdrawn (Figs 4.2 and 4.3). If a sheath with a

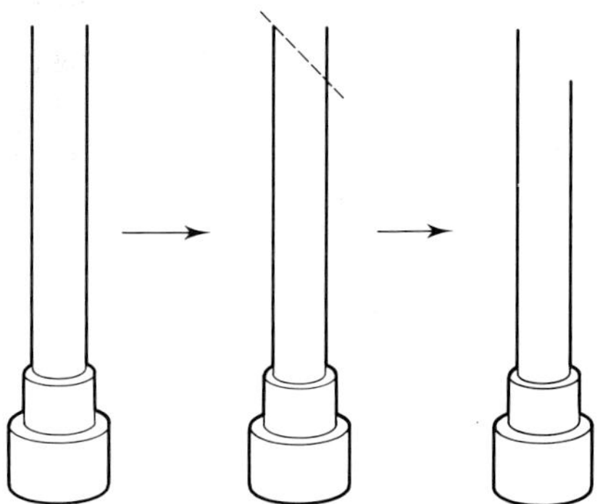

Fig. 4.2 The aperture of the introducer sheath is made larger by cutting it obliquely, making it easier to withdraw the foreign body into the sheath.

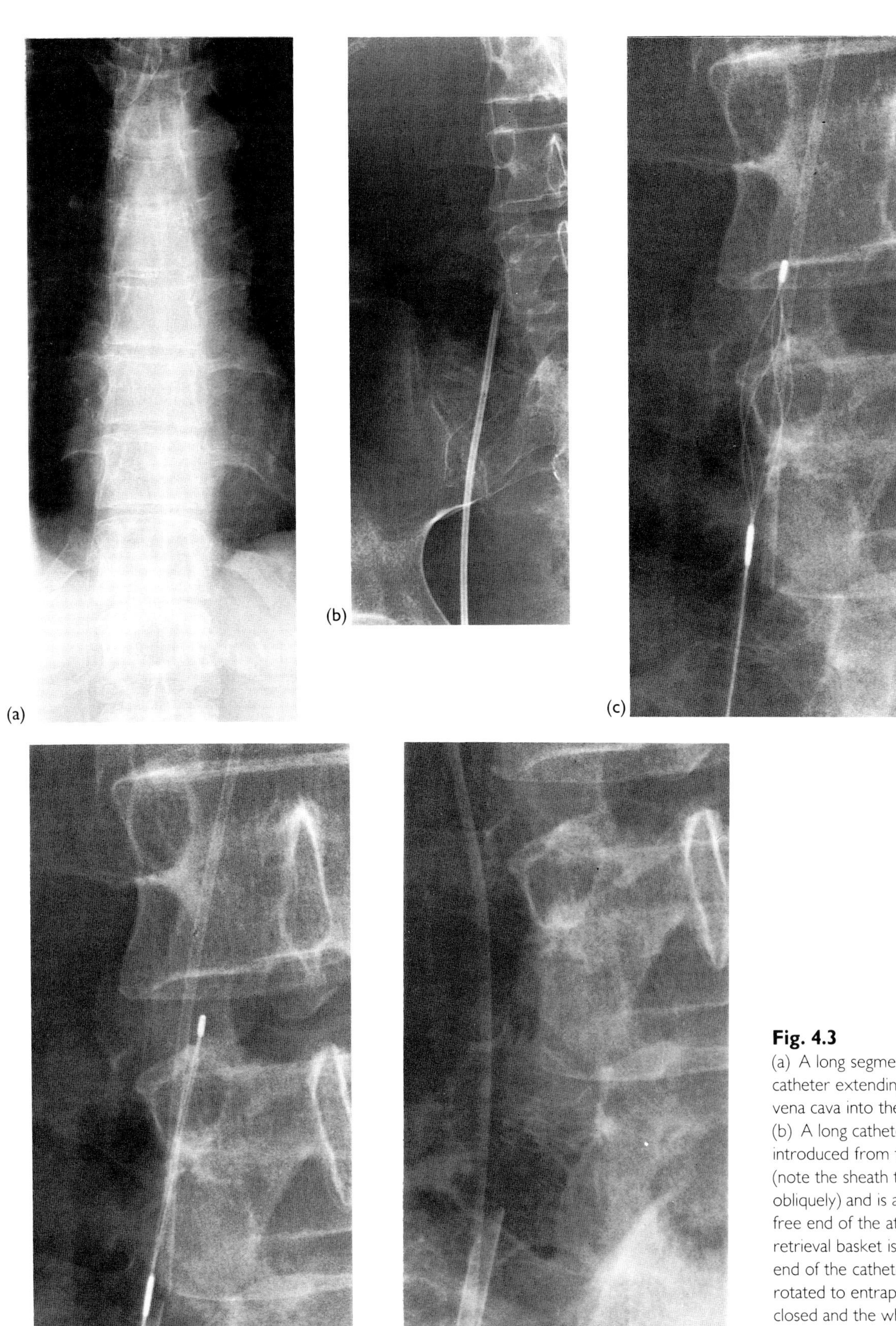

Fig. 4.3
(a) A long segment of an intravascular catheter extending from the superior vena cava into the inferior vena cava. (b) A long catheter sheath assembly is introduced from the right femoral vein (note the sheath tip has been cut obliquely) and is advanced close to the free end of the atheter fragment. (c) A retrieval basket is advanced to the free end of the catheter fragment and is rotated to entrap it. (d) The basket is closed and the whole assembly withdrawn. (e) The free end of the catheter is directed into the sheath during withdrawal. Reproduced with permission from Professor D.J. Allison.

detachable valve is used it may be possible to withdraw the foreign body without withdrawing the sheath. This is particularly helpful if there are several fragments or foreign bodies.

If the foreign body cannot be withdrawn into the sheath it may be necessary for a femoral cutdown to be performed.

Pigtail/preshaped catheters

The pigtail or other preshaped catheter used either alone or with a preshaped or tip deflecting guidewire can be used to dislodge or pull a foreign body into a more suitable position for retrieval and is particularly useful if a loose end is not available for snaring or if the catheter fragment is knotted (Fig. 4.4) (Thomas and Sievers, 1979). (If the knot is loose it may be reduced by simple traction of the pigtail catheter.) The tip of the catheter is used to engage the foreign body. It may take a great deal of patience but once the loop of the catheter is engaged the foreign body may be pulled with it. Once the foreign body has been pulled into a more convenient position, a loop snare or retrieval forceps can be used to grasp it. If the foreign body is wedged and the pigtail catheter cannot dislodge it, a loop snare can be used to snare the tip of the pigtail catheter (or any curved catheter for that matter) or a guidewire passed through the catheter and then the catheter and loop snare can be withdrawn simultaneously with more force (Fig. 4.5).

Loop snare *(Fig. 4.6)*

This can be made from equipment available in the angiography suite (Dotter *et al.*, 1971; Edwards and Sowton, 1978). A nontapered catheter (5–9 Fr) with a guidewire doubled back on itself to pass through the catheter and form a loop is all that is required. Guidewires of 0.018–0.021 inch diameter can be used in 7 Fr catheters (Fig. 4.6b).

There are several commercially available loop

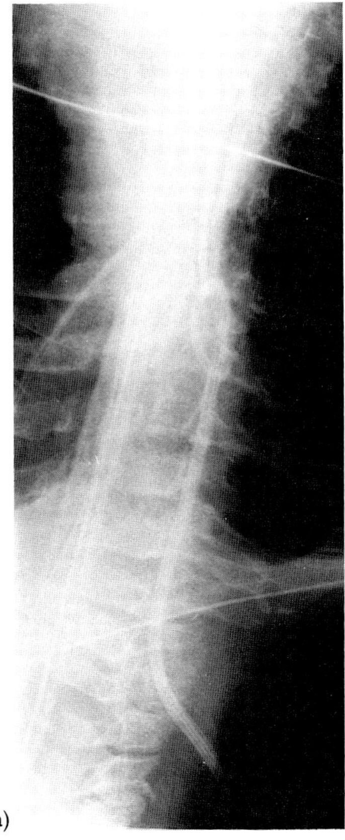

(a)

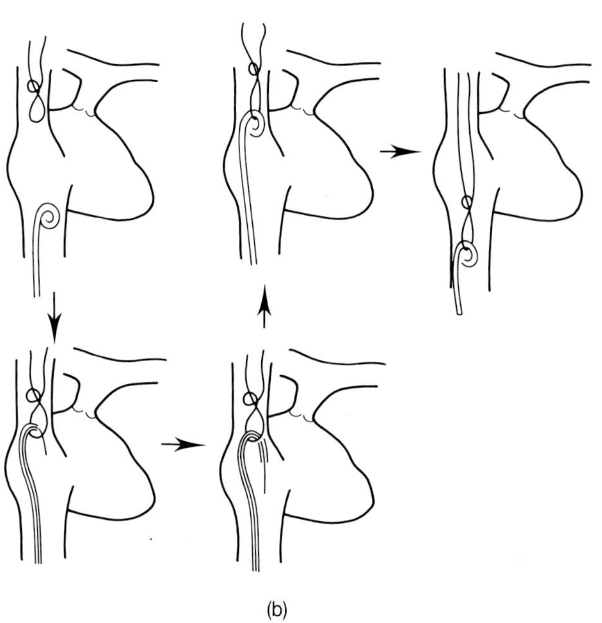

(b)

Figure 4.4
(a) A knotted Swan-Ganz catheter in the superior vena cava. (b) A pigtail catheter is used to dislodge the foreign body into a more suitable position for retrieval.

snares with angled wire loops (from 30° to 90°), varying loop diameters and materials (Yedlicka *et al.*, 1991) and an Allison fragment grasper which consists of a wire which will form a complete loop around fragments from 4 to 9 Fr in diameter. The tip is deflected by a three-ring plastic handle (Fig. 4.7). The catheters used may be straight or slightly angled. More recently a loop snare retrieval system

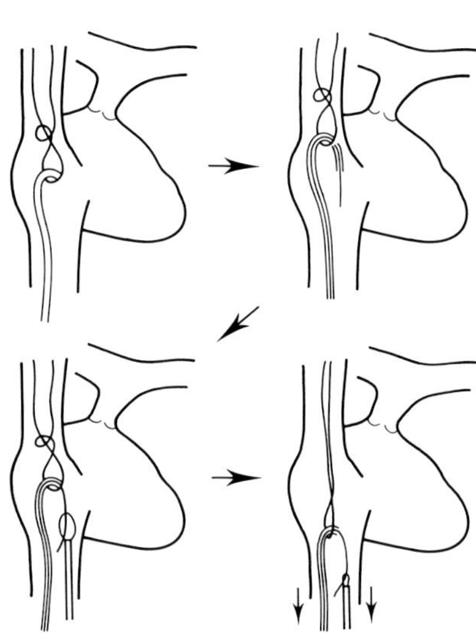

Fig. 4.5 If the pigtail catheter is unable to dislodge a wedged foreign body, a loop snare can also be used to snare the tip of the pigtail catheter or a guidewire. The foreign body can then be dislodged by withdrawing the catheter/wire and loop snare simultaneously.

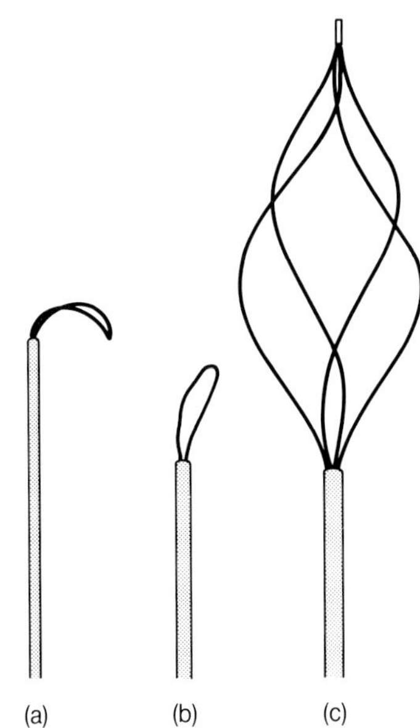

Fig. 4.6 Examples of loop snares and Dormia baskets
(a) Angled wire loop retriever. (b) 'Home-made' loop snare – nontapered catheter with 0.018 inch wire doubled back on itself. (c) Stainless steel helical loop basket.

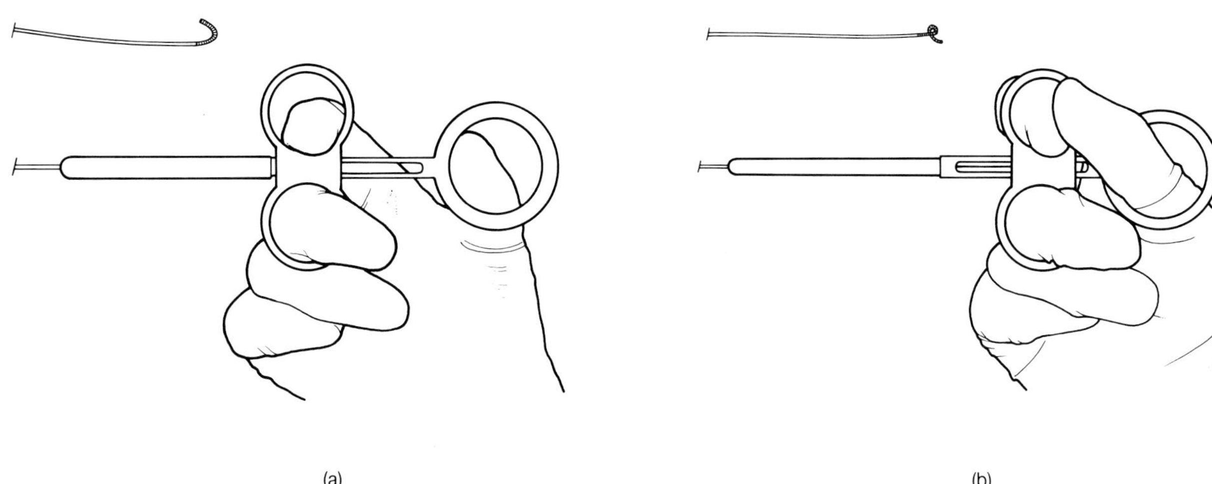

Fig. 4.7 Allison fragment grasper
(a) When the handle is in the relaxed position, the curve on the tip of the wire is open. (b) When the handle is activated the tip of the wire will form a complete loop.

with a three-lumen catheter has been described which allows a central guidewire to be maintained to allow repeated snaring and has fewer problems with blood loss and friction (Furui *et al.*, 1992).

The wire loop and catheter are advanced close to the foreign body and the wire loop is manipulated so that it passes over a loose end of the foreign body. The foreign body is then gripped by advancing the catheter and withdrawing the wire loop so that the foreign body is caught tightly between the two (Figs 4.8 and 4.9). The whole assembly is then gently withdrawn into the sheath at the puncture site.

Retrieval forceps

Grasping forceps can now be mounted on fairly flexible 3 Fr shafts and so can negotiate even tortuous vessels. The grasping portion can be of different types such as rat tooth- or alligator-type forceps. The device is passed through a large bore catheter, for example, 7 Fr, so it can be placed as close to the foreign body as possible. Care must be taken to use the forceps gently to avoid damage to the vessel wall. The grasping forceps are slowly opened and simultaneously advanced gently. The forceps are used to grasp one end of the foreign body and the forceps and guiding catheter are then withdrawn as a unit (Selby *et al.*, 1990). Although this is not recommended as the method of first choice, it can be useful in retrieving foreign bodies when other methods have failed, particularly when the foreign body does not have a free end.

Dormia basket *(Fig. 4.6c)*

This consists of a helical loop basket on a wire which passes through an 8 Fr catheter set. The basket is advanced up to the foreign body and the Dormia basket rotated so that the free end of the foreign body gets caught up in it. The foreign body is then snared by withdrawing the wire and basket towards the catheter or sheath, or advancing the catheter so that the foreign body is gripped firmly, the basket and catheter are then withdrawn into the introducer sheath. This system is useful for retrieving misplaced embolization coils (Chuang, 1979) (Fig. 4.10) or balloons, and fragments of angioplasty balloon as well as intravascular bullets.

Adjunctive retrieval methods

Adjunctive retrieval methods have been described (Uflacker *et al.*, 1986). The balloon retrieval method may be useful when fragments of wire or catheter have wedged peripherally in the pulmonary circulation. It involves passing a balloon catheter beyond the foreign body. The balloon is inflated gently and pulled back so dislodging the foreign body into a position more amenable to other retrieval techniques.

These retrieval methods can be used in arteries or veins and a femoral approach is the preferred method by most radiologists as manipulation from the groin is usually easier and more comfortable for the patient, and should a surgical cutdown be necessary this is better performed in the groin than the neck.

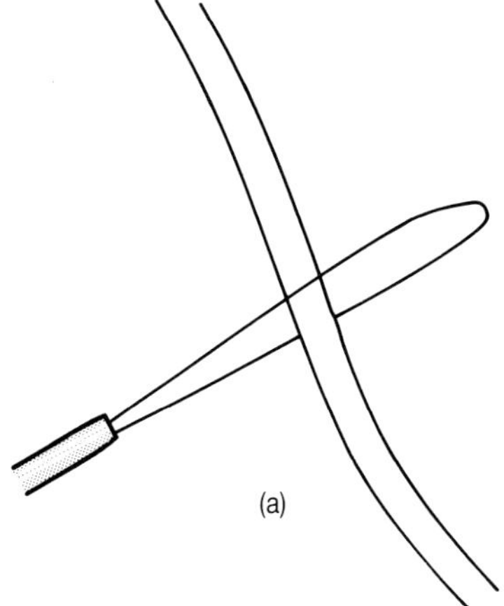

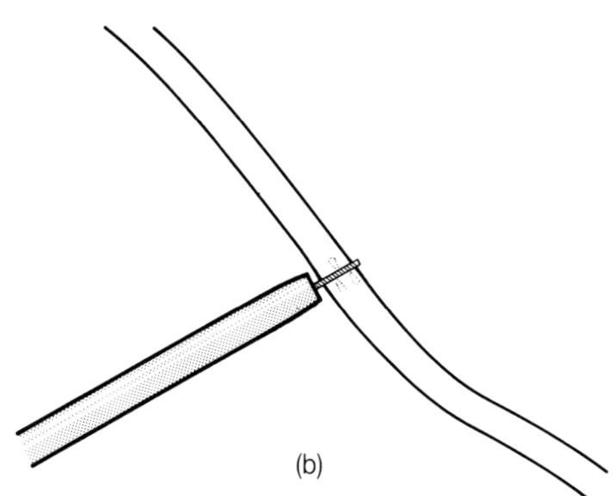

Fig. 4.8 Snaring a foreign body
(a) The wire loop is passed over the loose end of a foreign body so that the foreign body lies within the loop. (b) The foreign body is caught tightly by advancing the catheter and withdrawing the wire loop.

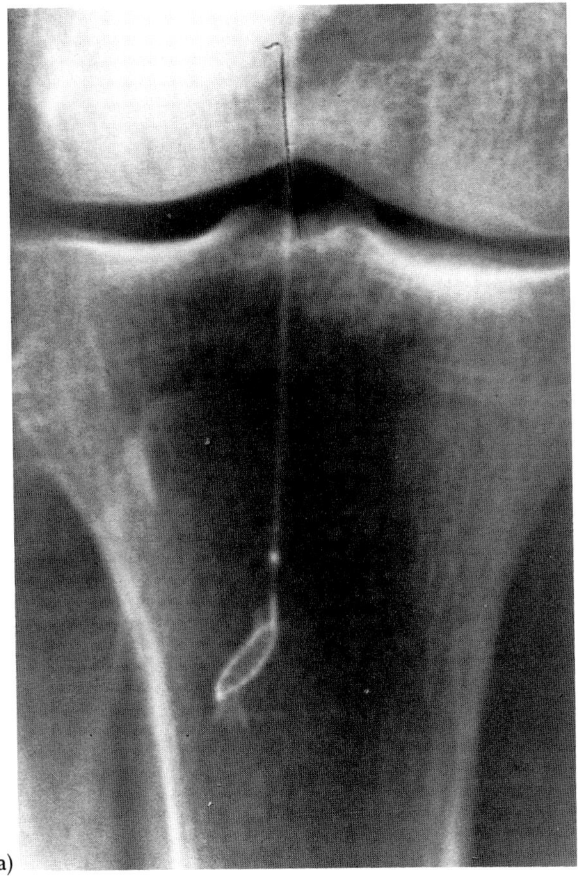
(a)

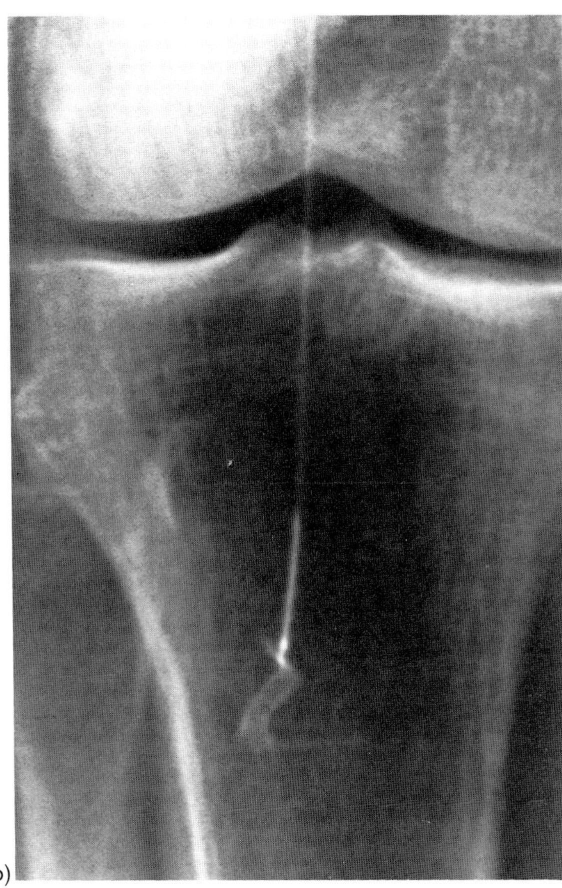
(b)

Fig. 4.9 A loop snare with a 90° angled wire loop (goose neck) is used to retrieve a misplaced coronary stent which has lodged in the distal popliteal artery.
(a) The snare has been positioned around the stent. (b) The proximal part of the stent is caught tightly in the snare and removed successfully through a 10.5 Fr sheath.

COMPLICATIONS (Table 4.4)

Failure

Unsuccessful attempts to retrieve foreign bodies are not strictly complications. Failure may be due to the use of an inappropriate retrieval method, for example, if there is no free end it will not be possible to use the loop snare, and a hook system, such as that described using a pigtail catheter, or forceps technique, will be necessary.

Fragmentation of foreign bodies

The foreign body may fragment further on retrieval, particularly if it has remained *in situ* for several months as catheter material tends to degrade. If a fragment does break off it may be possible to direct it to a site more amenable for retrieval. In the ascending aorta, for example, digital pressure on the carotid arteries will direct the fragment towards the lower limbs where it may lodge at the femoral bifurcation or be caught by digital pressure at the groin.

Complications
Failure
Further fragmentation of foreign body
Embolization of foreign body
Cardiac arrhythmia
Vascular perforation

Table 4.4 *Complications of percutaneous retrieval of intravascular foreign bodies*

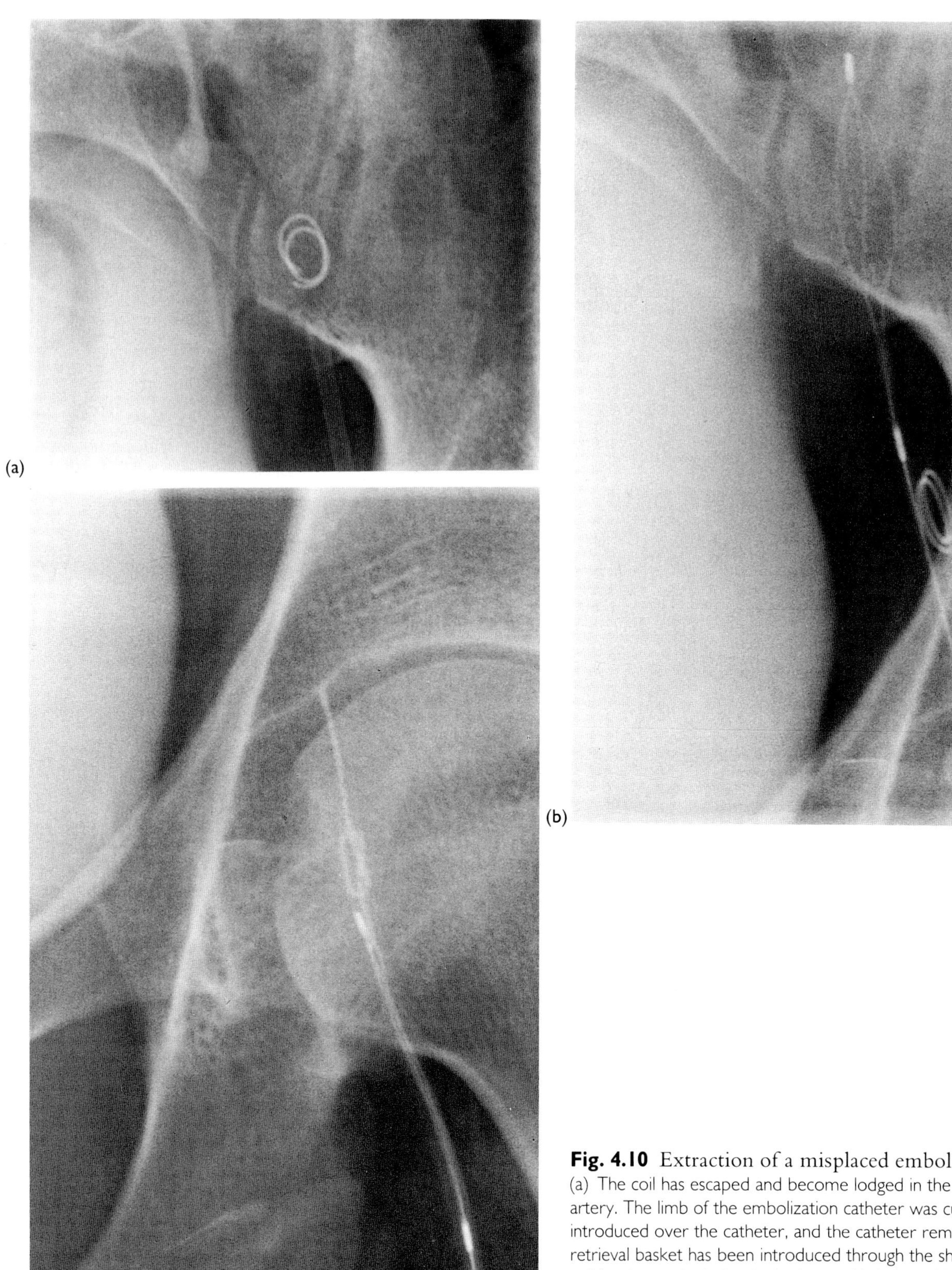

Fig. 4.10 Extraction of a misplaced embolization coil
(a) The coil has escaped and become lodged in the left external iliac artery. The limb of the embolization catheter was cut off, a sheath introduced over the catheter, and the catheter removed. (b) A retrieval basket has been introduced through the sheath. (c) The coil has been caught by the basket and is being withdrawn into the sheath. Reproduced with permission from Grainger and Allison, *Diagnostic Radiology* Churchill Livingstone.

Embolization of foreign bodies

In an attempt to snare a foreign body it may be dislodged and embolize further peripherally making percutanoues retrieval more difficult.

Cardiac arrhythmias

This is probably the most common complication when retrieving foreign bodies from the right side of the heart and pulmonary artery. Usually this is transient, but cardiac monitoring is essential throughout the procedure and pharmacological control of arrhythmias may prove necessary.

Vascular perforation

This risk is increased when using rigid instruments such as the Dormia basket or a rigid forceps in tortuous vessels. When using a home-made loop snare the point of the loop may be sharp or rigid and this could cause perforation. If part of the foreign body is thought to be extravascular it may be prudent to avoid percutaneous retrieval and resort to surgical retrieval. This may depend on the site of the foreign body, for example, if venous, percutaneous retrieval may be safe.

POSTPROCEDURAL CARE

This does not differ from the postprocedural care of any other angiographic procedure.

The patient should be put on bed rest for 12–24 hours, depending on whether a venous or arterial puncture was necessary and the size of sheath that was used. The groin, blood pressure and pulse should be monitored regularly to detect any haemorrhage following the procedure. Antibiotic therapy is not necessary unless septicaemia is present, or the risk of septicaemia is considered high. It may be appropriate to send the foreign body for culture.

REFERENCES

CHUANG VP (1979) Non-operative retrieval of Gianturco coils from abdominal aorta. *American Journal of Roentgenology* **132:** 996–7.

DAVIES J, ALVARES R, ALLISON DJ (1981) An intra-cardiac foreign body: diagnosed non-invasively and removed non-surgically. *British Journal of Radiology* **54:** 987–9.

DOERING RB, STENMER EH, CONNOLLY JE (1967) Complications of indwelling venous catheter with particular reference to catheter emboli. *American Journal of Roentgenology* **11(4):** 259–66.

DOTTER CT, ROSCH J, BILBAO M (1971) Transluminal extraction of catheter and guide fragments from the heart and great vessels. *American Journal of Roentgenology* **111:** 467–72.

EDWARDS AC, SOWTON E (1978) Management of embolised central venous catheters. *British Medical Journal* **77:** 669–70.

FISHER RG, FERREYRO R (1978) Evaluation of current techniques for nonsurgical removal of intravascular iatrogenic foreign bodies. *American Journal of Roentgenology* **130:** 541–8.

FURUI S, YANAUCHI T, MAKITA K, TAKESHITA K, IRIE T, TSUCHIYA K, SAWADA S, NAKAMURA H, OKAZAKI M (1992) Intravascular foreign bodies: Loop-snare retrieval system with a three-lumen catheter. *Radiology* **182:** 283–4.

HIEBERT CA, GREGORY FJ (1974) Bullett embolism from the head to the heart. *Journal of the American Medical Association* **229:** 442.

MEYERS L (1945) Intravenous catheterization. *American Journal of Nursing* **45:** 930.

RUBINSTEIN ZJ, MORAG B, ITZCHAT Y (1982) Percutaneous removal of intravascular foreign bodies. *Cardiovascular and Interventional Radiology* **5:** 64–8.

SELBY JB, TEGTMEYER CJ, BITTNER GM (1990) Experience with new retrieval forceps for foreign body removal in the vascular, urinary and biliary systems. *Radiology* **176:** 535–8.

THOMAS HA, SIEVERS RE (1979) Non-surgical reduction of arterial catheter knots. *American Journal of Radiology* **132:** 1018–19.

THOMAS J, SINCLAIR-SMITH B, BLOOMFIELD D AND DAVACHI A (1964) Nonsurgical retrieval of a broken segment of steel spring guide from the right atrium and inferior vena cava. *Circulation* **30:** 106–8.

TURNER DP, SOMMERS BG (1954) Accidental passage of polyethylene catheter from cubital vein to right atrium. *New England Journal of Medicine* **251:** 744–45.

UFLACKER R, LIMA S, MELICHAR AC (1986) Intravascular foreign bodies: percutaneous retrieval. *Radiology* **160:** 731–5.

VUJIC I, MOORE L, MCINLEY RE (1986) Retrieval of coil after unintentional embolization of ileocolic artery. *Radiology* **160:** 563–4.

YEDLICKA JW JR, CARLSON JE, HUNTER DW, CASTANEDA-ZUNIGA WR, AMPLATZ K (1991) Nitinol Gooseneck Snare for removal of foreign bodies: Experimental study and clinical evaluation. *Radiology* **178:** 691–3.

Suggested further reading

UFLACKER R (1991) Percutaneous retrieval of intravascular foreign bodies. In: *Current Practice of Interventional Radiology*, pp. 121–6. Edited by Kadir S. BC Decker Inc., Philadelphia.

KADIR S, ATHANASOULIS CA (1982) Percutaneous retrieval of intravascular foreign bodies. In: *Interventional Radiology*, pp. 379–90. Edited by Athanasoulis CA, Pfister RC, Greene RE, Roberson GH. WB Saunders, Philadelphia.

ROSSI P, PAVONE P (1990) Foreign body retrieval. In: *Interventional Radiology*, pp. 717–27. Edited by Dondelinger RF, Rossi P, Kurdziel JC, Wallace S. Theme Medical Publishers, New York.

ALLISON DJ (1991) Vascular extraction techniques (i) Intravascular foreign bodies. In: *Diagnostic Radiology: An Anglo-American Textbook of Organ Imaging*, pp. 2351–2. Edited by Grainger RG, Allison DJ. Churchill Livingstone, Edinburgh.

CHAPTER 5

Inferior vena cava filters

Irving Wells

Indications for inferior vena cava filtration 94

Indications for suprarenal inferior vena cava filtration 95

Contraindications to inferior vena cava filtration 95

Preparation for inferior vena cava filtration 96

Technique of insertion: general considerations 97

Materials: filter types and their insertion 100

Technical problems: how to overcome them 112

Complications of inferior vena cava filtration 114

Results and choice of filter 115

References 116

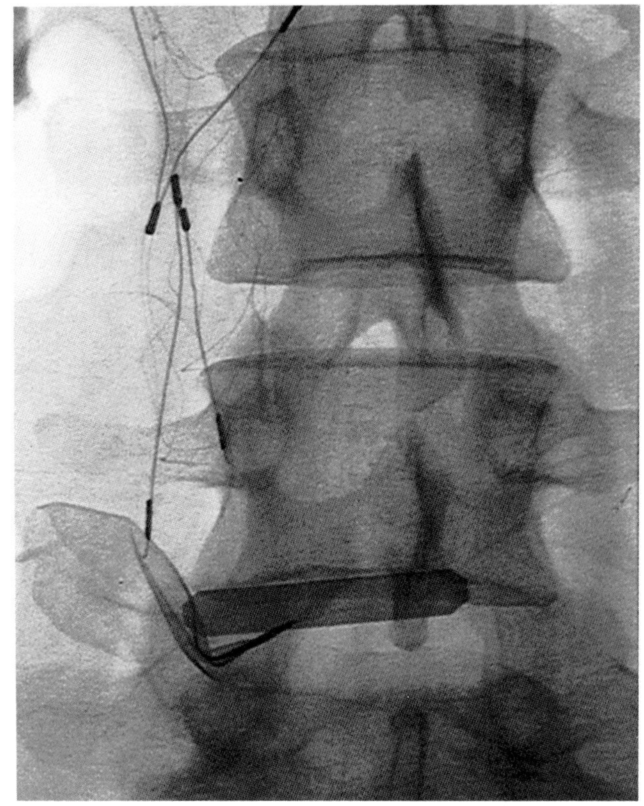

INDICATIONS FOR INFERIOR VENA CAVA FILTRATION

The sudden, unexpected death from pulmonary embolism is unpreventable. Mortality from pulmonary embolism in patients known to be at high risk is however largely preventable if suitable therapeutic measures are taken. Patients who survive an episode of embolism form the highest risk group and have about a 30% chance of dying from further embolism if they remain untreated (Dalen and Alpert, 1975). Ever since the work of Barritt and Jordan (1960) anticoagulation has formed the mainstay of this treatment but as many as 18% of patients so treated may have further emboli (Santos and Lansman, 1982) and 8% may die as a result (Dalen and Alpert, 1975). Inferior vena cava (IVC) filtration can reduce recurrence to an average of 2% with only a small number of these proving fatal.

The clearest and best established indications for IVC filtration are therefore in those patients who are thought to have had a pulmonary embolus but who cannot be anticoagulated and those who are having recurrent embolism despite adequate anticoagulation (Wells, 1989). These and other possible indications are listed in Table 5.1.

Indications
Pulmonary embolism with contraindication to anticoagulation
Recurrent pulmonary embolism despite adequate anticoagulation
Deep venous thrombosis with contraindication to anticoagulation
Deep venous thrombosis in patients with pre-existing pulmonary hypertension
Failure of existing filter device
Postpulmonary embolectomy

Table 5.1 *Indications for inferior vena cava filtration*

If a patient conforms to any of the categories listed in Table 5.1, IVC filtration should be considered. More detailed consideration will now be given to each of these categories.

Pulmonary embolism with contraindication to anticoagulation

Every effort should be made to be as sure as possible about the diagnosis of embolism. This will usually be by ventilation/perfusion scanning and the demonstration of a source of embolism. Although pulmonary angiography is the gold standard it is not usually necessary to perform it as a preliminary to filter insertion in most cases.

Contraindications to anticoagulation would comprise:

1 Recent gastrointestinal, brain or urinary tract bleeding
2 Trauma, including surgery within the last few days
3 Pre-existing clotting disorders

If the risks of anticoagulation are thought to be greater than filter insertion in a patient with pulmonary embolism then filtration must be performed as a matter of urgency.

Recurrent pulmonary embolism despite adequate anticoagulation

It is essential to confirm that anticoagulation has been adequate by checking that indices of clotting have been in the therapeutic range. Recurrence of embolism may be suggested by clinical signs but should be reinforced by a technetium-99m-labelled macroaggregate isotope perfusion scan which shows fresh defects compared with the initial ventilation/perfusion scan. If these criteria are met filter insertion should be performed urgently and anticoagulation should be continued.

Deep venous thrombosis with a contraindication to anticoagulation

The exact risk of pulmonary embolism from untreated deep venous thrombosis is not known. The risk of embolism from thrombus confined to the infrapopliteal segment is not known but is thought to be extremely low. Some authors, however, have documented pulmonary embolism in as many as 10% of patients with iliofemoral thrombosis despite anticoagulation (Scott Norris *et al.*, 1985). If the thrombus is shown to be free floating on venography this has been associated with a risk of embolism as high as 60% (Scott Norris *et al.*, 1985). It is therefore recommended that all patients with nonadherent iliofemoral thrombus who cannot be anticoagulated undergo insertion of an IVC filter. Patients with more distal or adherent thrombus should be closely followed up and should have an isotope perfusion scan after a week or so. Any evidence of embolism should lead to urgent filter insertion.

Deep venous thrombosis in patients with pre-existing pulmonary hypertension

Patients with reduced pulmonary vascular reserve are much less able to tolerate pulmonary embolism. Any patient known to have pulmonary arterial hypertension who develops a deep venous thrombosis should be considered for filter insertion in addition to anticoagulation. The age of the patient and the cause of the pulmonary hypertension will need to be taken into consideration in deciding if filter insertion is appropriate. Elderly patients with chronic lung disease as a cause of their loss of pulmonary vascular reserve might be considered less suitable for filter insertion than younger patients with other causes of pulmonary arterial hypertension for example.

Failure of existing filter device

If a filter has been placed wrongly, for example, in an iliac vein, a second filter may be placed above to ensure adequate caval filtration. Rarely, propagation of thrombus proximally from an occluded filter may necessitate the placement of a second filter via the jugular route.

Postpulmonary embolectomy

This operation is rarely performed in the UK but American authors indicate that filtration should be performed at the time of embolectomy to protect against the high risk of subsequent embolism (Mansour *et al.*, 1985). The level of risk is not clear however.

INDICATIONS FOR SUPRARENAL INFERIOR VENA CAVA FILTRATION

The positioning of filters will be discussed in detail in a later section on 'Technique of insertion', but there are occasional reasons for positioning a filter in the IVC above the renal veins (Orsini and Jarrel, 1984) rather than in the conventional infrarenal position. These are listed in Table 5.2. If recurrent emboli occur despite filter insertion below the renal veins a further filter above the renal veins may prevent recurrence of embolism if this is due to poor clot trapping efficiency of the first filter or if the emboli are arising from the renal or gonadal veins.

Thrombus in the infrarenal IVC
Renal vein thrombosis causing emboli
Renal transplant patient requiring filtration
Recurrent emboli despite infrarenal IVC filtration

Table 5.2 *Indications for suprarenal inferior vena cava (IVC) filtration*

CONTRAINDICATIONS TO INFERIOR VENA CAVA FILTRATION

These are listed in Table 5.3 and will be considered in more detail below.

An inability to gain access to the cava because there is thrombosis both above and below the renal veins or because thrombosis has extended all the way up the cava would make insertion of a filter impossible.

Severe bleeding disorder may be a relative contraindication because of bleeding from the puncture site.

The risks of irradiation to the foetus should be taken into consideration in the first 20 weeks of pregnancy but these must be balanced against the risks of pulmonary embolism in the mother.

The life expectancy of filters is largely unknown. The new designs, although rigorously tested, have not yet stood the test of time in patients. This fact should be born in mind when considering filtration in younger patients with a long life expectancy. Because of the clinical setting in which filtration is considered however doubts about the longevity of filters should not stand in the way of offering this

No access to IVC
Bleeding disorder
Pregnancy
Large diameter of IVC

Table 5.3 *Contraindications to inferior vena cava (IVC) filtration*

potentially life-saving treatment, but using a filter with as much previously documented safe usage as possible would certainly be prudent. This is discussed further in the section on 'Choice of filters'.

A caval diameter of greater than 28 mm is a contraindication to using some types of filter; this will be considered further in the section on 'Materials'.

PREPARATION FOR INFERIOR VENA CAVA FILTRATION

Regardless of what type of filter is used and what the indication for filter insertion is, there are some essential preprocedural requirements. These are listed in Table 5.4 and will be considered in detail below.

Obtain full informed patient consent
Check patency of inferior vena cava and iliofemoral veins
Decide on filter type to be used
Decide on site of insertion
Check patient clotting parameters

Table 5.4 *Preprocedure requirements*

Full patient consent

There are risks attached to the insertion of filters both at the time of filtration and in the longer term. These have to be balanced against the risks of the patient's thromboembolic disease and this equation of risks should be discussed as fully as possible with the patient. When using the newer filter types it is particularly important to stress that the risk of longer term complications cannot be fully known.

Check patency of inferior vena cava and iliofemoral veins

The presence of clot in these veins may already have been determined in diagnosing the embolic disease, but if not it is essential to assess their patency to allow a decision to be made about the site of insertion, which may in turn influence the type of filter chosen. Doppler ultrasound has been shown to be useful in this context (Baxter *et al.*, 1990) and is to be recommended as a noninvasive way of imaging the veins.

Decide on filter type to be used

The section on 'Materials' later in this chapter will give some detail of the different types of filter currently available. As firm a decision as possible should be made on what type of filter to use as this will have a bearing on the way the procedure is set up and will ensure that both the patient and the radiologist will be able to anticipate how the procedure will be conducted.

Decide on site of insertion

Filters are designed to be inserted either from the femoral or the internal jugular veins (Coleman, 1986). McCowan *et al.* (1990) have described the use of the external jugular vein but this is unlikely to be used commonly because of its small size and tortuous course. Most vascular radiologists are much more familiar with the technique of femoral vein catheterization than jugular vein and would prefer to use this route. Clearly this would depend upon the iliac veins and lower IVC being free of clot. The right iliac veins run a straighter course than the left and some filter types are not recommended for use from the left femoral vein. The right femoral vein will therefore be the entry point of choice in most patients with the right internal jugular and left femoral in reserve. Distribution of thrombus, type of filter being used and operator experience will determine which of the reserves will be used.

Check patient clotting parameters

Patients undergoing filtration will often have been anticoagulated and it is desirable to return the clotting parameters towards normal for the time of the filtration. Patients who have been heparinized can usually be dealt with just by stopping the heparin 4 hours before the procedure. Patients on Warfarin will usually require fresh frozen plasma to cover the procedure as there will not usually be time to allow the effects of the Warfarin to wear off. In practice a prothrombin ratio of up to 1.5 is acceptable and patients with clotting times in the therapeutic range should not be excluded as long as the possibility of reversal is anticipated if bleeding cannot be controlled at the end of the procedure. Guidance from haematology colleagues should be sought in more complex disorders of clotting.

Technique of insertion: general considerations

There are many aspects of the technique of filter insertion which are common to all types of filter and these will be dealt with now. Specific points of insertion of individual filters will be dealt with in the next section.

In common with all angiographic techniques the percutaneous insertion of caval filters is best carried out in the radiology department where facilities for X-ray screening and the back-up of trained technicians and nurses will be optimal. It should be a sterile procedure and sedative premedication may be required. Routine premedication is not however necessary.

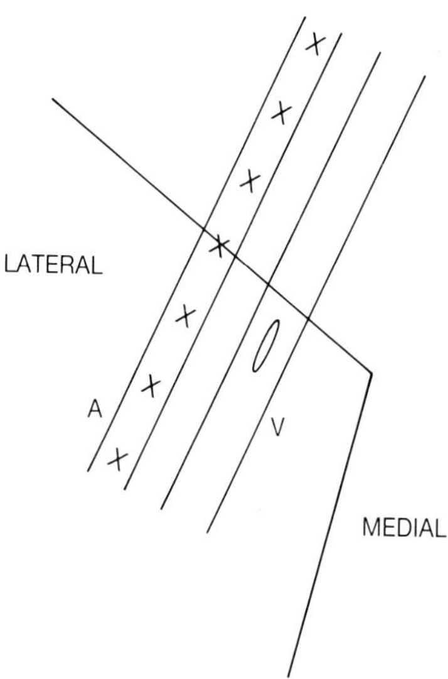

Fig. 5.1 Marking the skin along the arterial pulse will give a visual guide to the position of the vein medial to it.

Femoral vein cannulation

Since the advent of cross-sectional imaging not all radiologists have had the opportunity to become experienced in femoral vein puncture and the following points may help to make the procedure as easy as possible. The femoral vein runs medial to the artery (Fig. 5.1) and marking the skin over the length of the femoral artery pulsation allows the radiologist to have a guide to the position of the artery which may facilitate venous puncture and help to prevent inadvertent arterial puncture. The artery should be palpated along its length during puncture which should be about 0.5 cm medial to it. A conventional Seldinger technique (Seldinger, 1953) is used, and a single anterior wall puncture is preferred.

The pressure within the vein is lower than that within the artery and not always sufficient to produce a good flow of blood through the needle. Venous pressure can be elevated by the Valsalva manoeuvre and if the patient is able to cooperate it may be found helpful to ask the patient to perform this manoeuvre during the cannulation. Also, because of the relatively low pressure within the vein, suction on the needle via a connecting tube and syringe will help to produce a flow of venous blood confirming position within the vein.

Once the vein has been entered a guidewire is passed to the proximal common iliac vein so that a cavogram can be performed. Guidewires are often included with filter sets but any angiographic wire of suitable size may be used. A J wire may show less tendency than a straight wire to inadvertently enter the ascending lumbar veins and a size of 0.038 inches is usually adequate. If there is any doubt that the iliac veins are free of clot the wire should only be introduced a little way into the vein and insertion of a dilator will then allow injection of contrast medium to show the lumen of the veins before passing the wire up to the cava.

Internal jugular vein puncture

This technique is not commonly practiced by most radiologists who would usually select the femoral route by choice. However, when the iliac or femoral veins are occluded or when there is clot in the lower cava it will be necessary to use the jugular and the following points may be found helpful.

The right internal jugular should be used because the passage from it into the superior vena cava is much less tortuous than on the left (Fig. 5.2a and b). The usual sterile precautions should be taken and the patient is placed in the supine position with the head

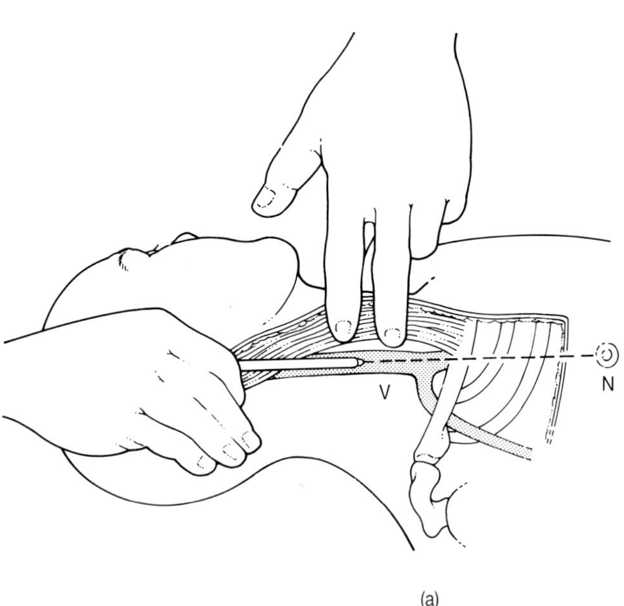

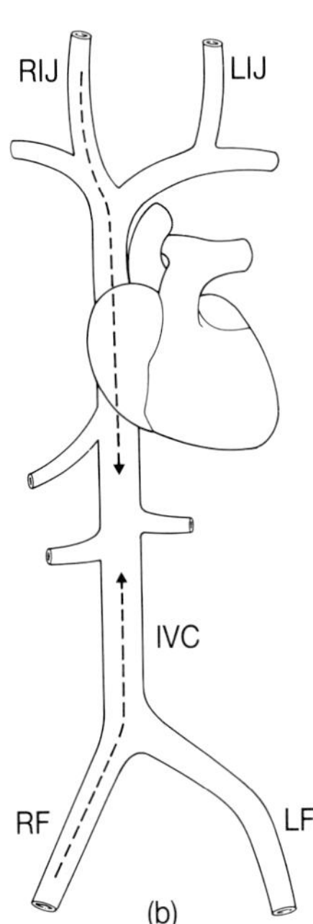

Fig. 5.2
Puncture of the right femoral (RF) or right internal jugular vein (RIJ) will allow a much less tortuous course to the inferior vena cava (IVC) than the left femoral (LF) or left internal jugular vein (LIH).

turned to the left. The vein runs along the lateral border of the sternocleidomastoid muscle about 1.5 cm lateral to the carotid artery. Palpation of the arterial pulse with the left hand will help to prevent inadvertent carotid puncture. The puncture should be made in the line of the vein with the needle angled towards the feet at the midpoint of the neck. As with the femoral vein suction on the needle and performance of the Valsalva manoeuvre by the patient may facilitate the flow of venous blood and help entry into the vein. Once in the vein, passage of a wire and catheter to the distal inferior vena cava for performance of a cavogram is usually straightforward. Occasionally the wire will enter an hepatic or even a renal vein however and passage of the wire and position of the catheter for the cavogram must be checked with fluoroscopy. If the wire does tend to enter the hepatic vein a catheter with a gentle curve can be used to guide it through the cava (Fig. 5.3).

Fig. 5.3
A curved catheter may be useful for directing the guidewire away from the heaptic vein (HV) when approaching the inferior vena cava (IVC) from the jugular vein.

Inferior vena cavogram

This should be performed in all cases prior to filter insertion. The diagnostic catheter should be positioned at the lower end of the cava just above the iliac vein confluence. A pigtail or straight flush catheter may be used and 40 to 50 ml contrast medium should be injected at a rate of 15 to 20 ml/sec. Films should be obtained at a rate of 2 films/sec for 5 sec. Digital subtraction may modify these figures. Table 5.5 lists the information required from the cavogram and this will now be considered in detail.

Check caval patency
Check position of renal veins
Check for normal anatomy
Measure diameter of cava

Table 5.5 *Reasons for performing inferior vena cavogram*

Patency of the cava should have been determined prior to the procedure but the cavogram reaffirms this and rules out any extension of the clot since the preprocedural workup.

Determining the position of the renal veins is vital if the filter is to be properly positioned below them. They usually lie at the level of L1 to L2. Using the technique of cavography outlined above the renal veins will usually be identifiable by the negative contrast of unopacified blood entering the cava from them. Despite the large number of veins entering the cava it is usually only the renal and hepatic veins which show up in this way (Fig. 5.4). Before the cavogram is performed metal markers should be taped to the patient to allow easy location of the position of the renal veins and the iliac venous confluence during subsequent positioning of the filter.

Congenital variations in the normal anatomy of the IVC and its tributaries occur sufficiently commonly that the radiologist inserting filters should be aware of them and check for them in each case (Chuang *et al.*, 1974). The most important include left-sided IVC (0.2%), duplication of the IVC (2.2%) and left retroaortic renal vein (3%). Duplication of the IVC may require the placement of two filters or a single filter in the common suprarenal cava. A left-sided cava is best approached from the jugular or left femoral route (Sardi and Minken, 1987). The left renal vein normally passes in front of the aorta before draining into the IVC. A retroaortic vein passes behind the aorta and usually drains into the IVC at a lower level than usual and sometimes

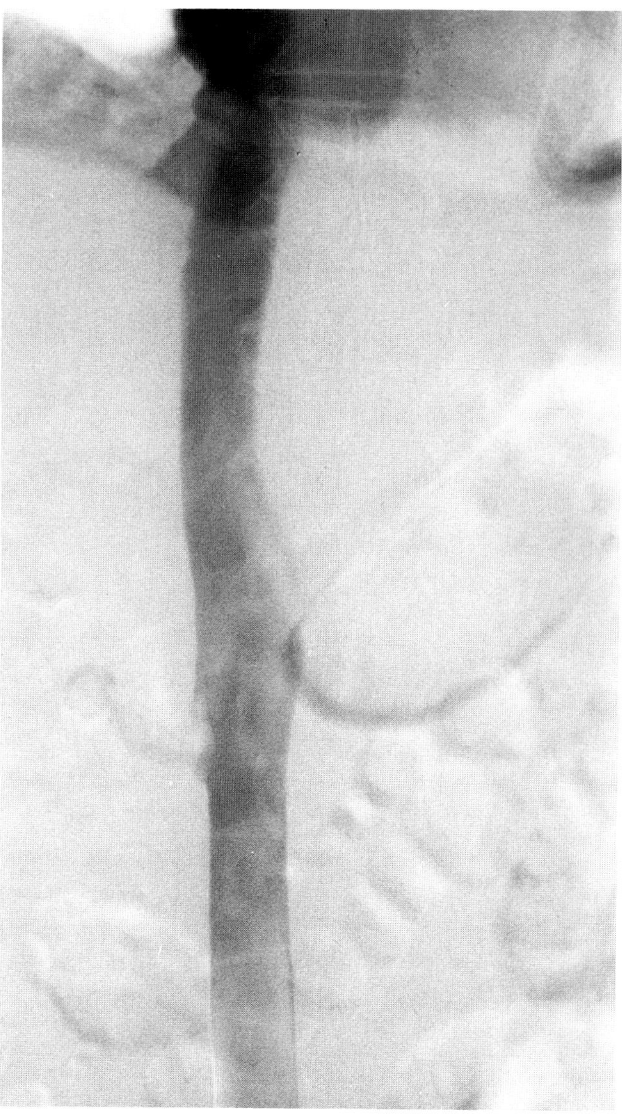

Fig. 5.4
Inferior vena cavogram showing position of renal and hepatic veins by entry of unopacified blood represented as filling defects in the contrast column.

into the iliac vein. If this anomaly is found it may be necessary to position the filter above the renal veins. If abnormal anatomy of the renal veins is suspected on the cavogram or if they are not clearly identified it will be helpful to explore for them by using a visceral catheter and injecting contrast medium into individual veins as required.

The diameter of the IVC should be measured in all cases before filter placement. Some filters are not recommended for use in cavas with a diameter of greater than 28 mm. These megacavas are thought to occur in about 3% of the population (Prince *et al.*, 1983). The diameter should be measured where the filter will be placed, usually half-way between the iliac confluence and the renal veins. The magnification factor on a cavogram is about 23% which means that the radiograph measurement should be multiplied by 0.77. Caval diameter may be increased in patients with elevated right atrial pressure and is influenced by respiration and the Valsalva manoeuvre. If a filter that is too small for the diameter of the IVC is used there is a risk of migration, conversely if the filter is too large it will not open properly which may hinder clot trapping efficiency.

Postinsertion routine

After the filter has been placed haemostasis must be achieved by compression of the puncture site for 10 min. A longer time may be required if clotting is still deranged.

An abdominal film should be obtained to document the position of the filter.

Patients should stay on bed rest for at least 12 hours and should have the standard postangiography observations of pulse, blood pressure and puncture site.

If not contraindicated anticoagulation with heparin may be restarted after 6 hours.

In the long-term follow-up of patients with filters abdominal films in anteroposterior (AP) and lateral projections will be useful to check filter stability and integrity, and ultrasound is useful to check IVC patency. Both computed tomography (CT) and magnetic resonance imaging (MRI) have been used in the follow-up of filters and may be useful in demonstrating clot in or adjacent to the filter.

MATERIALS: FILTER TYPES AND THEIR INSERTION

With the introduction of the Mobin-Uddin umbrella in 1970 (Fig. 5.5) it soon became clear that the placement of filters via the venous system represented a considerable advance over surgical caval plication not least because the operative mortality was reduced from 10 to 0.3% (Santos and Lansman, 1982). There were problems with migration and thrombosis with the umbrella however and it was only with the arrival of the Kimray-Greenfield (KG) filter in 1973 (Greenfield *et al.*, 1973) that filtration began to gain widespread acceptance. The KG filter (Fig. 5.6a and b) achieved a caval patency of 95% in follow-up over periods as long as 10 years and migration was only very rarely reported (Mansour *et al.*, 1985). The recurrent embolism rate was reduced to less than 2%. These figures bear comparison with all the newer filters and there has been more experience worldwide with the KG filter than with any other. The disadvantage of the KG filter however was that it was delivered through a 24 Fr capsule which required cutdown onto the jugular vein for access, a technique usually performed by surgeons. Descriptions have been made of a modified Seldinger approach to the placement of KG filters involving dilation of a 24 Fr track into the femoral or jugular veins using renal dilators (Denny *et al.*, 1985). This heroic intervention has however been rendered largely redundant by the development of a new generation of filters which can be introduced through 12 or even 10 Fr catheters thus allowing a normal Seldinger approach to be used. It is these filters which are clearly in the preserve of the radiologist which will now be described in detail.

Fig. 5.5 Diagram of Mobin-Uddin umbrella.
Note the large surface area of the filter which leads to a high thrombosis rate.

Table 5.6 lists some of these new filter types and each will now be considered individually.

Titanium Greenfield filter (12 French)
Cardial filter (12 French)
Bird's Nest filter (10 French)
LGM (Venatech) filter (12 French)

Table 5.6 *New generation of inferior vena cava filters currently available in the UK*

Titanium Greenfield filter

SPECIFICATION

This filter is manufactured by Medi-Tech of Watertown, Massachusetts and currently costs £787 in the UK. It is a modified version of the stainless steel KG filter being designed for delivery through a 12 Fr introducer. The filter (Fig. 5.7) is made of titanium

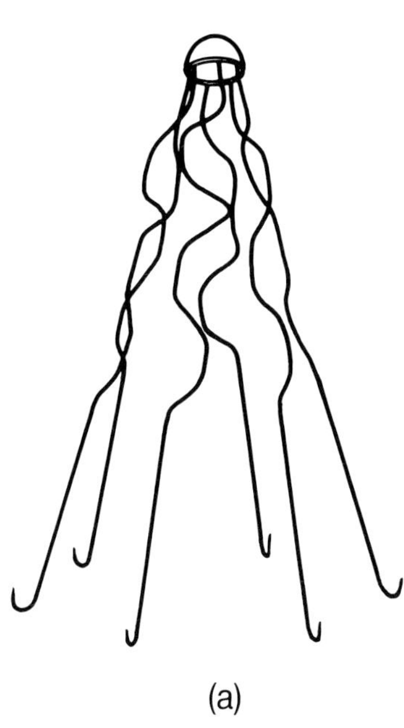

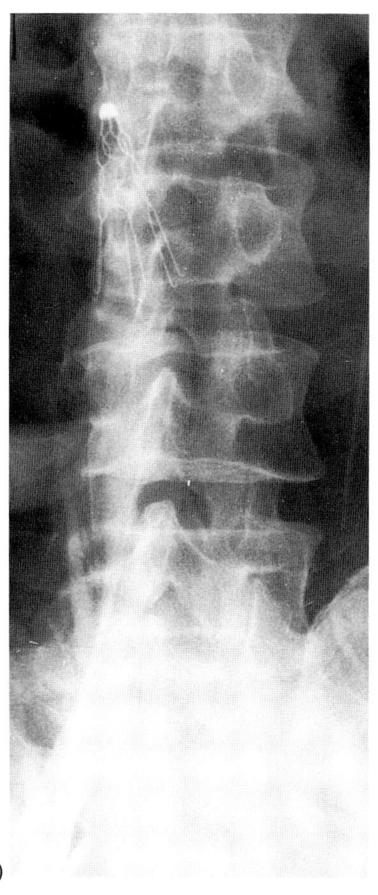

Fig. 5.6
(a) Diagram of the Kimray-Greenfield (KG) filter which is made up of six sprung legs with terminal hooks. (b) Inferior vena cavogram showing KG in position but with slight tilt.

which has a high strength-to-weight ratio and is both nonthrombogenic and corrosion resistant. In common with the KG filter the legs of the titanium filter (TGF) are loaded into a delivery capsule so that when the filter is extruded from its capsule the legs spring apart and the hooks on the end of the legs stick firmly into the wall of the IVC. This type of design leads to firm anchorage but may cause damage to the caval wall. Early versions of the TGF indeed did cause caval perforation (Teitelbaum *et al.*, 1989) and this has led to a modification of the hook design which has rectified this problem.

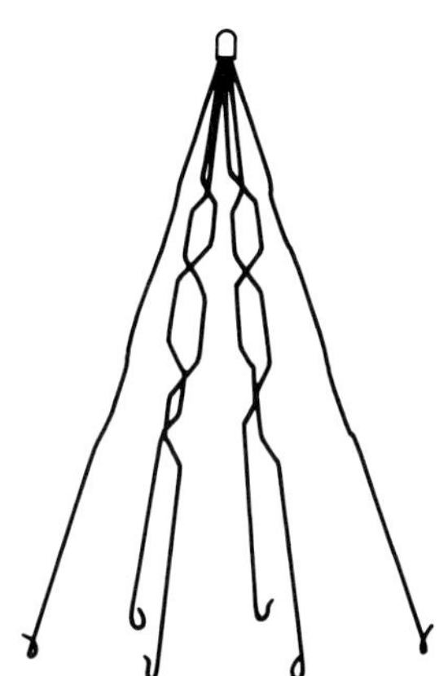

Fig. 5.7 Diagram of the Titanium Greenfield filter (TGF).
It is wider at its base than the Kimray-Greenfield filter but hook penetration is limited by the shape of the hooks.

INSERTION TECHNIQUE

The most important advice to any radiologist considering the placement of an IVC filter is to read and follow the manufacturer's instructions.

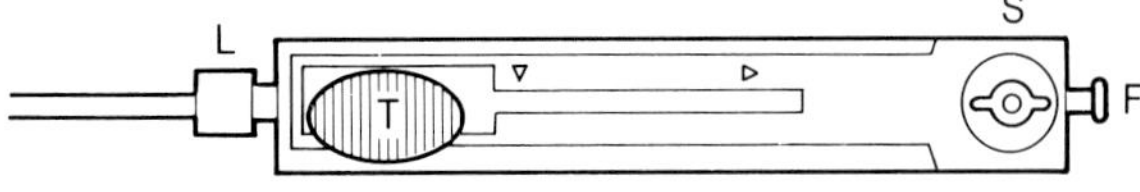

Fig. 5.8 View from above the Titanium Greenfield filter handle.
The colour coded release tab (T) is slid back towards the stopcock (S) to release the filter. Note also the luer lock (L) for connection to the sheath and the flush port (F).

The filter set comprises an introducer catheter with a preloaded filter and filter release handle (Fig. 5.8) and a 12 Fr sheath/dilator set. The design of these is quite different depending on whether the femoral or jugular route has been selected, and colour coding of the release tab on the handle and of the sheath/dilator has been used to prevent mixing of sets and to alert the user to the need to use the appropriate set (green for femoral and blue for jugular). If the wrong set is used the filter will be placed upside down (Fig. 5.9).

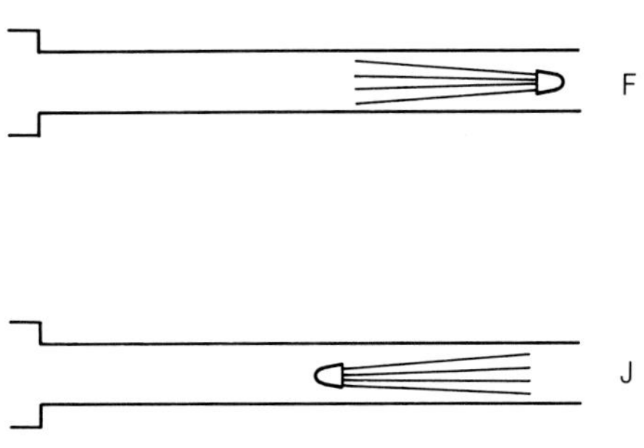

Fig. 5.9
The Titanium Greenfield filter is preloaded in opposite orientations for femoral (F) or jugular (J) insertion and the sets are not interchangeable.

To insert a femoral TGF the following steps will be required:

1 Insert a 0.038 inch guidewire and catheter and perform a cavogram. A TGF is not recommended for use in cavas of greater than 28 mm diameter.
2 Determine the level of the desired filter position below the renal veins and relate it to skin markers (Fig. 5.10).

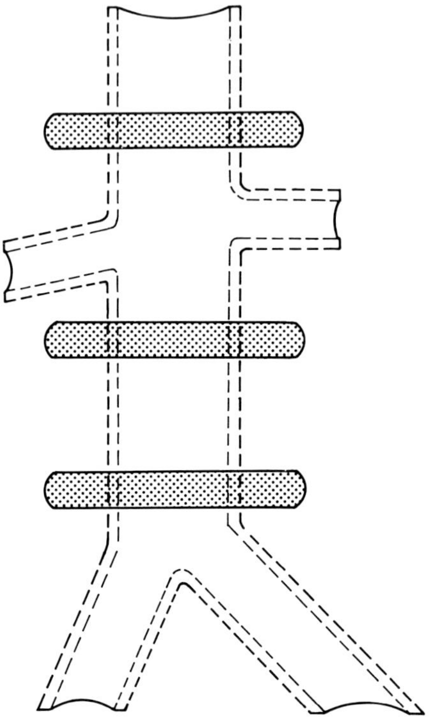

Fig. 5.10
Radio-opaque skin markers, in this case three ampoule files, attached prior to the cavogram will give a useful guide to filter positioning during subsequent fluoroscopy. A filter should be positioned above the lowest marker with its top at the level of the middle marker.

3 Reinsert the guidewire to a level above the renal veins and replace the cavogram catheter with the 12 Fr sheath/dilator. Vascular dilators may be used to facilitate entry of sheath/dilator and note that the sheath/dilator should not be advanced except over the guidewire.
4 When the sheath is above the renal veins the dilator and guidewire are removed together and the sheath flush adaptor which is attached to the sheath is screwed onto the sheath to allow it to be flushed with heparinized saline (Fig. 5.11).
5 The filter introduction set should be filled with

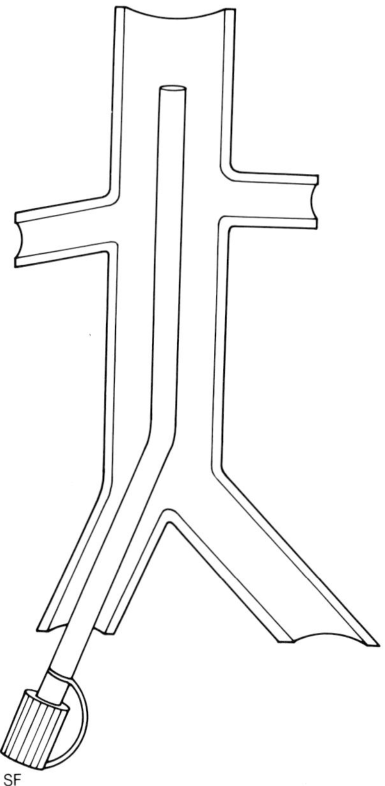

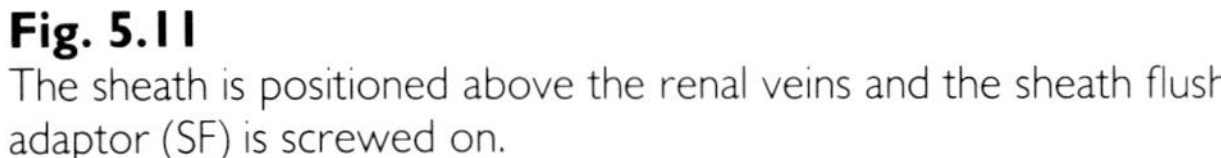
Fig. 5.11
The sheath is positioned above the renal veins and the sheath flush adaptor (SF) is screwed on.

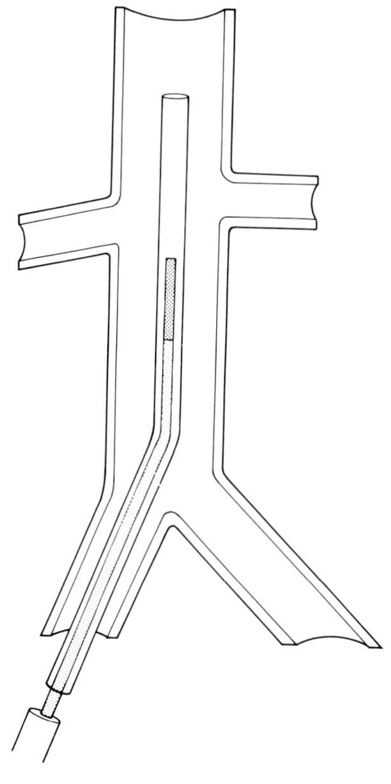

Fig. 5.12 Top of filter capsule should be just below renal veins inside sheath.

heparinized saline to prevent the risk of clot forming around the filter.

6 The sheath flush adaptor is now removed from the end of the sheath and the filter introducer catheter is inserted through the sheath until the top of the carrier capsule is at the desired level for the top of the filter (Fig. 5.12). The carrier capsule is easily visualized and should not be advanced beyond the sheath. If the sheath has been properly positioned above the renal veins this will not occur.
7 Holding the filter introducer catheter in position, the sheath is now withdrawn over it until the luer lock connection at its hub can be screwed to the handle of the introducer catheter (Fig. 5.13).
8 Make final adjustments to the position of the filter, then remove the paper band from the introducer handle and slide the green tab to the left unlock position and then slowly all the way back to the proximal end of the track. This will release the filter. Do not pull or push on the introducer catheter while releasing the filter as this will change the position of filter release.
9 Remove the filter introducer catheter through the sheath. A cavogram can now be performed through the sheath if required but this is not usually necessary. The sheath is then also removed (Fig. 5.14).

To insert a jugular TGF the steps required are similar to those for a femoral filter but with the following important differences:

1 Having performed a cavogram the sheath/dilator is advanced over the wire to a position about 6 cm below the renal veins (Fig. 5.15).
2 After removal of the dilator and wire the introducer catheter is advanced through the sheath so that the carrier capsule lies with its proximal end just below the renal veins (Fig. 5.16).
3 The sheath is then withdrawn over the introducer catheter and the filter released as with the femoral method.

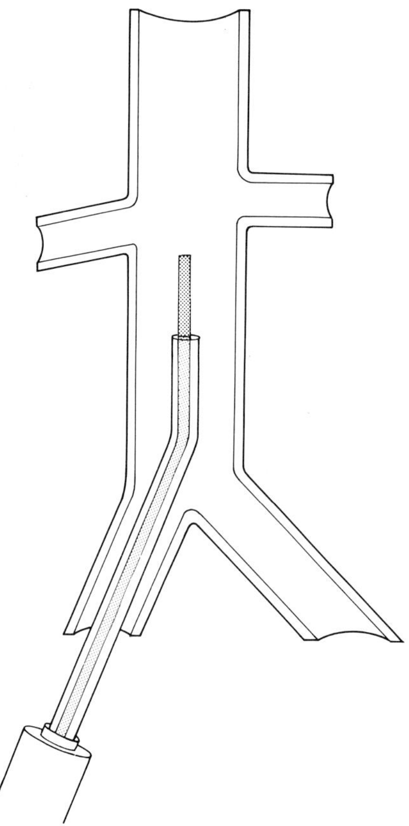

Fig. 5.13 The sheath is screwed to the luer connector on the release handle exposing the filter capsule. Note that the filter capsule has not moved.

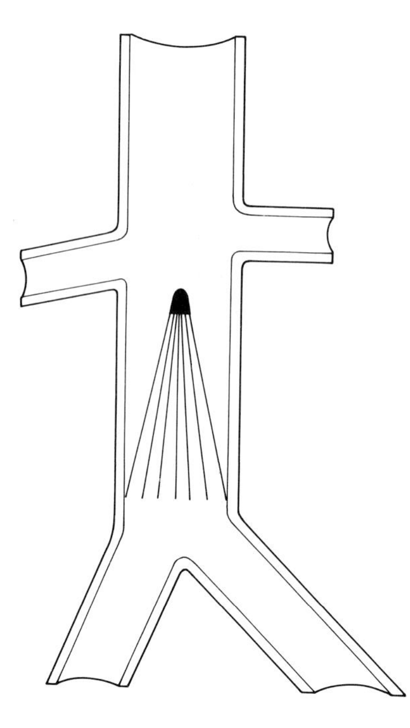

Fig. 5.14 After release the filter should lie between the renal veins and the iliac confluence in an untilted position.

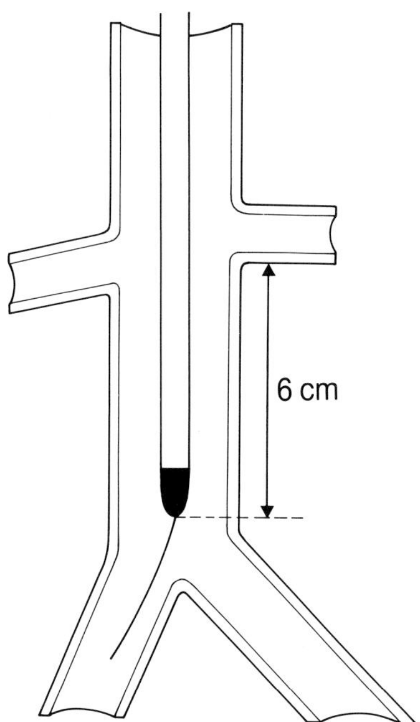

Fig. 5.15 From the jugular the sheath is advanced at least 6 cm below the renal veins.

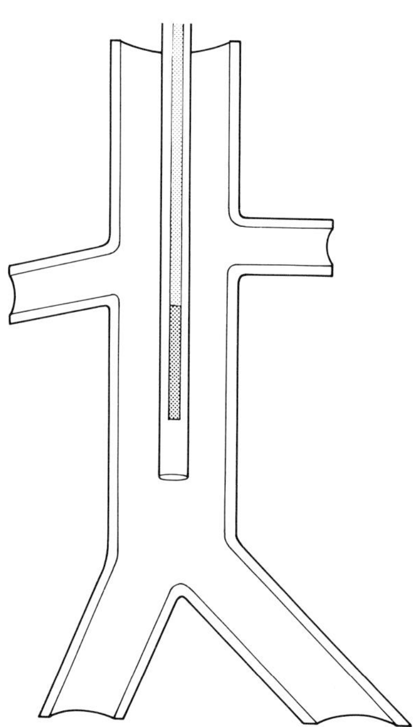

Fig. 5.16 Proximal end of carrier capsule lies just below renal veins.

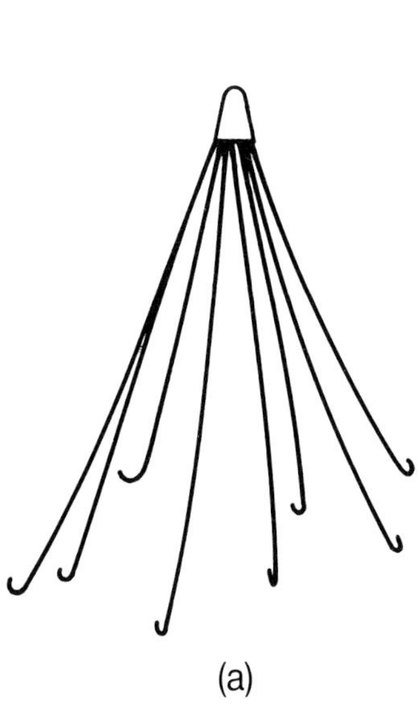

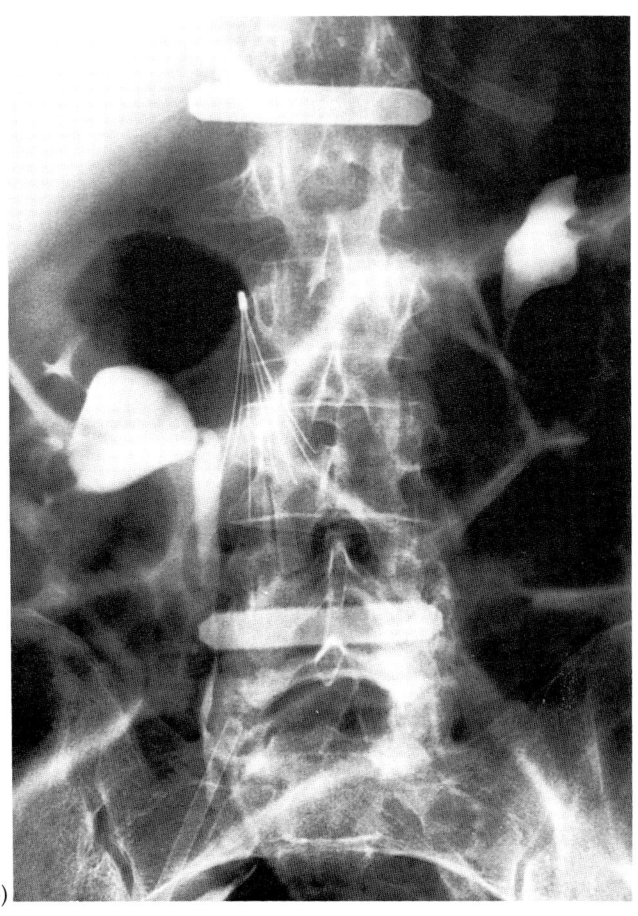

Fig. 5.17
(a) The Cardial filter is similar to the Titanium Greenfield filter but has eight straight legs. (b) Radiograph of Cardial filter. Note the ampoule file markers.

Cardial vena cava filter

SPECIFICATION

This filter is manufactured by Cardial in France and distributed in the UK by BVM Medical Ltd at a cost of £590. The Cardial filter is made of stainless steel and is of a similar general type to the Greenfield filter, being sprung loaded into a delivery capsule which is inserted via a sheath (Fig. 5.17a and b). The delivery mechanism is different and involves a lockable sprung wire. The Cardial filter is simple to place but some problems with sheath kinking and inadvertent filter extrusion in the sheath have led to recent design modifications. This will be considered further in the section on 'Technical problems'.

INSERTION TECHNIQUE

As previously stated it is essential to follow the manufacturer's instructions which are usually very comprehensive. Unfortunately the instructions with the Cardial are not detailed and include no diagrams. The set comprises a needle, J wire, sheath/dilator and preloaded filter insertion catheter. Both the sheath and the filter catheter contain side ports for flushing and the sheath has a haemostatic valve at its proximal end (Fig. 5:18). Separate sets are available for femoral and jugular placement and are not interchangeable.

To insert a Cardial filter from the femoral vein the following steps will be required:

1 Insert a 0.038 inch wire and catheter and perform a cavogram. No guidance is given about caval size limits for using the filter but it would be wise not to place it if the caval diameter is greater than 28 mm.
2 Determine the level of the desired filter position below the renal veins and relate it to skin markers (Fig. 5.10).
3 Reinsert the guidewire and replace the diagnostic catheter with the sheath/dilator positioning sheath about 6 cm below renal veins (Fig. 5.19). Keep

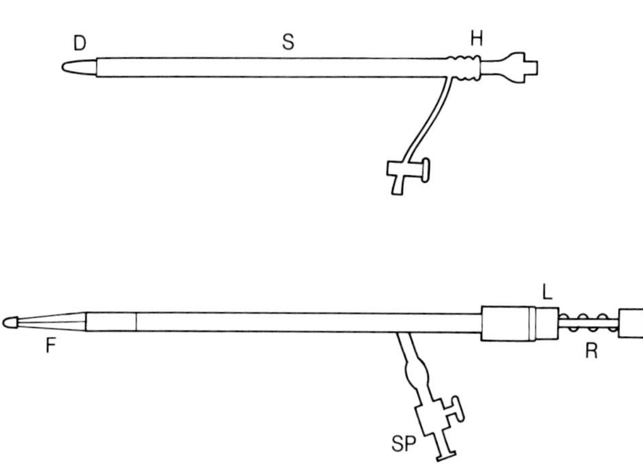

Fig. 5.18
The Cardial sheath (S) has a haemostatic valve (H). The introducer catheter has a preloaded filter (F), a side port for flushing (SP) and a lockable (L) sprung loaded release wire (R). D = dilator.

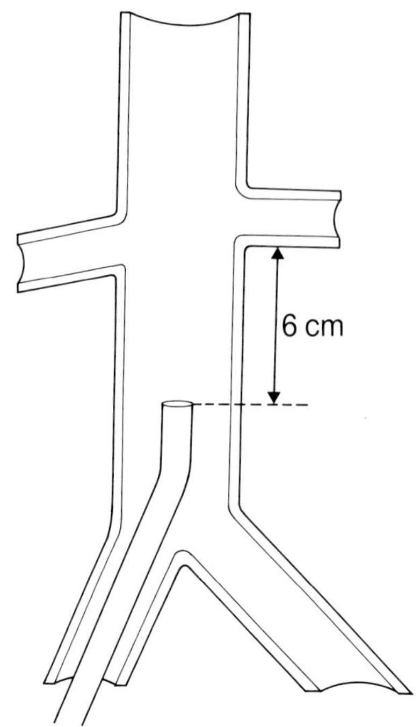

Fig. 5.19
Position sheath 6 cm below renal veins.

sheath flushed with heparinized saline.

4 Check that the locking screw on the filter catheter is fully turned in a clockwise direction. This will prevent the filter release wire being inadvertently pushed in during the procedure and accidentally releasing the filter during manipulation.
5 Carefully release the plastic cover from the filter by grasping the red end of the cover. Any disturbance of the filter's preloaded configuration will prevent proper placement. Flush the catheter with heparinized saline.
6 Pass the filter catheter through the sheath via the haemostatic valve until it can be advanced no further. The proximal end of the filter will emerge from the sheath and should lie just below the renal veins. Adjust its position if necessary by moving the sheath and filter catheter as a unit (Fig. 5.20).
7 To release the filter the locking screw is turned anticlockwise to unlock the release wire. Then the filter catheter is pulled slowly back against the firmly held proximal end of the catheter. It is important to understand that the end of the catheter is not pushed in to push the filter out but

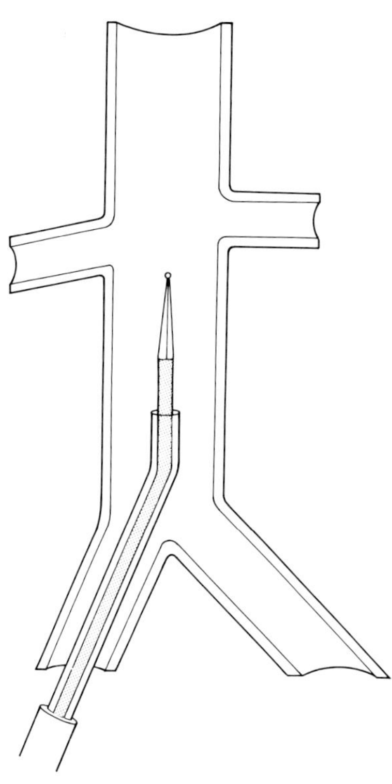

Fig. 5.20 Once the filter catheter is fully through the sheath, the filter can be finally positioned by moving the catheter and sheath together. Note it is important to keep the catheter fully through the sheath because unlike the Titanium Greenfield filter there is no mechanism to lock the sheath hub to the catheter.

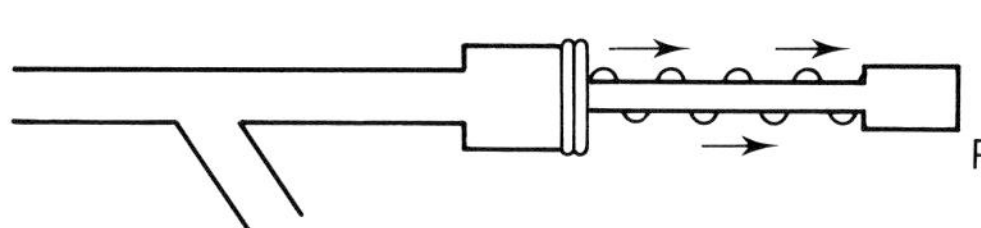

Fig. 5.21
To release the filter hold the plunger (P) still and pull the catheter back towards it in the direction of the arrows.

that the filter release wire is kept static while the catheter is pulled back along it (Fig. 5.21).

8 Remove the filter catheter through the sheath and then remove the sheath.

To insert a Cardial filter from the internal jugular vein follow the steps as for a femoral placement but with these differences:

1 Position the sheath at the level of the renal veins.
2 When the filter catheter is fully passed through the sheath the carrier capsule should be positioned with its proximal end just below the renal veins (Fig. 5.22).
3 Release the filter by pulling back against the end of the catheter as with the femoral method.

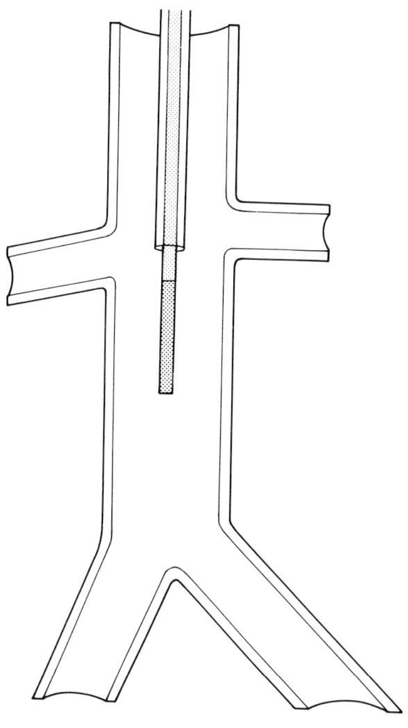

Fig. 5.22
Jugular Cardial carrier capsule is positioned just below the renal veins.

Gianturco-Roehm Bird's Nest vena cava filter

SPECIFICATION

The Bird's Nest filter (BNF) is manufactured by Cook Incorporated of Bloomington Indiana and costs £420 in the UK. It is designed for introduction through a 12 Fr sheath and is quite unlike any other filter (Roehm *et al.*, 1984). It consists of two stiff securing struts between which the operator forms a mesh of wire (Fig. 5.23a and b). It is made of stainless steel and although the securing struts are easily seen on fluoroscopy the fine mesh wires of the filter are not easily seen. Following early reports of filter migration (Roehm *et al.*, 1988), the securing struts were modified thus rectifying the problem.

INSERTION TECHNIQUE

The sequence of steps required to place the BNF is complex at first sight and the manufacturer's excellent instructions should be followed. A VHS video is available from them to help explain the technique. The set comprises a 12 Fr sheath/dilator and a preloaded filter catheter with a wire pusher and a side port for flushing (Fig. 5.24). Separate sets are available for femoral and jugular routes, the main difference being in the length of the sheaths and filter catheters.

To insert a femoral BNF the following steps will be required:

1 Insert a 0.038 inch wire and catheter and perform a cavogram. The BNF is suitable for use in all sizes of cava.
2 Determine the level of the desired filter position and relate it to skin markers (Fig. 5.10).
3 Replace the diagnostic catheter with the sheath/dilator and advance them to the hilt. Remove dilator and wire.
4 Flush the filter catheter with contrast medium and insert it through the sheath until the sheath hub can be luer locked to the filter catheter. The filter catheter will extend 15 mm beyond the tip of the sheath and its tip should be positioned just below the renal veins (Fig. 5.25).
5 Loosen the locking haemostatic valve and while

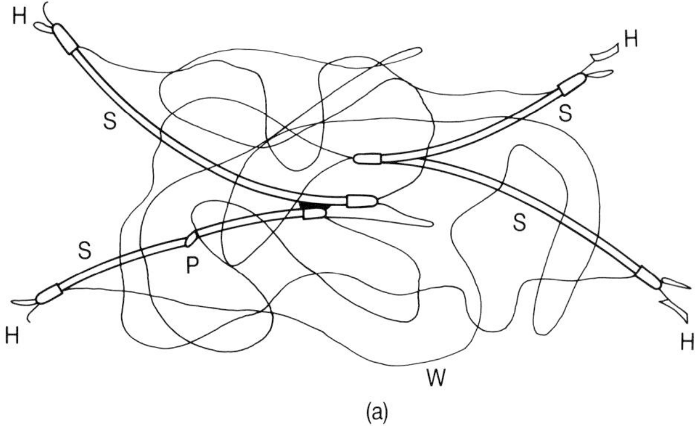

(b)

Fig. 5.23
(a) Diagram of Bird's Nest filter showing the fixation struts (S) with their terminal hooks (H). The wire pusher is attached to the struts at point (P) and the wire (W) is arranged around the struts.
(b) Radiograph of Bird's Nest filter. Note that the junction points of the struts overlap and that the wire is barely visible.

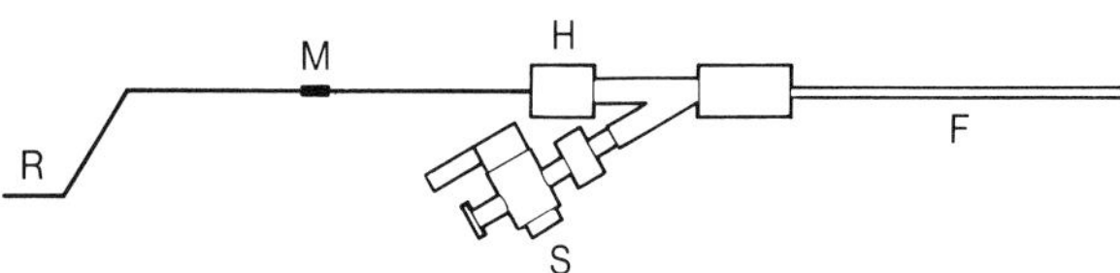

Fig. 5.24
The filter introducer comprises a preloaded filter (F) and a stopcock for flushing (S). Note that the wire pusher passes through a locking haemostatic valve (H) and has a mark to show position of catheter withdrawal (M) and a right angle handle (R).

holding the wire pusher still withdraw the catheter over it until reaching the mark on the wire. This will extrude the distal struts but leave their junction point just within the catheter (Fig. 5.26).

6 Advance the filter catheter forward gently 1–3 mm to push the strut hooks into the caval wall.
7 Making sure that the catheter remains in the IVC withdraw it 1–3 cm over the wire pusher making room for filter mesh formation.
8 To make the filter mesh the catheter is held still while the wire pusher is smoothly and slowly advanced through it until the junction point of the proximal struts is seen near the catheter tip (Fig. 5.27). This should be clearly seen fluoroscopically. The manufacturers state that some herniation of the mesh above the distal struts is of no consequence.
9 The catheter/sheath is now advanced until the junction point of the proximal struts is at the same level or just above the junction point of the distal struts (Fig. 5.28).
10 While holding the wire pusher still the catheter/sheath is withdrawn over it until the right angle handle of the pusher is reached. This extrudes the proximal struts and their hooks. The hooks are anchored by gently moving the pusher backwards and forwards (Fig. 5.29).

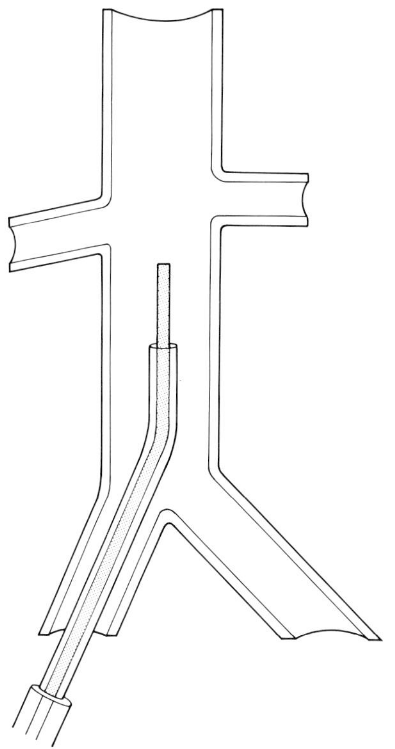

Fig. 5.25 Tip of filter catheter should be just below renal veins and it should be luer locked to the hub of the sheath.

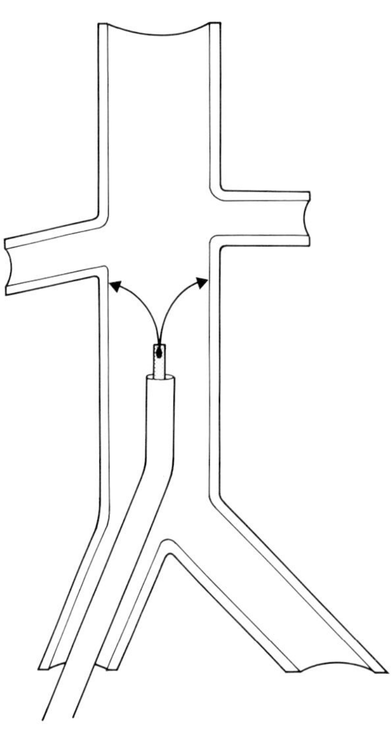

Fig. 5.26 When catheter is withdrawn to mark on wire pusher upper securing struts will be extruded but their junction point will still be just inside catheter.

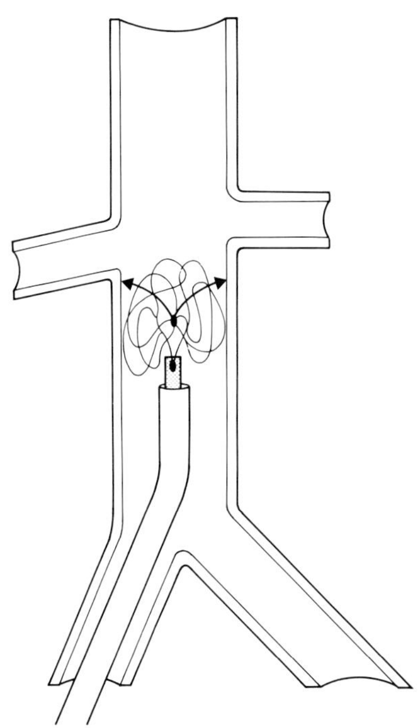

Fig. 5.27 The wire mesh of the filter is pushed out until the junction point of lower struts is seen near the tip of the catheter.

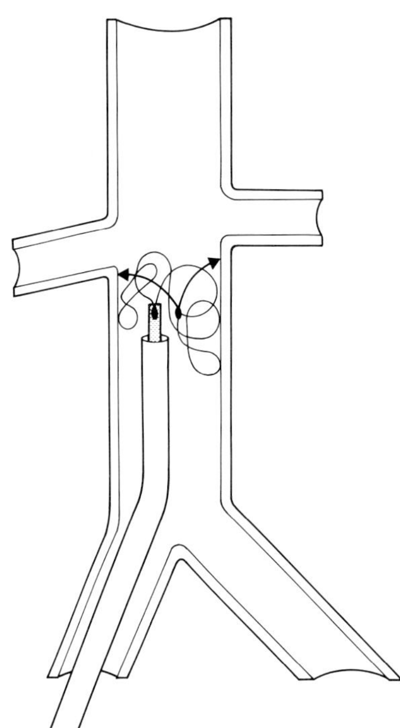

Fig. 5.28 The catheter is advanced until the junction point of the lower struts is at the same level or just above the junction point of the upper struts.

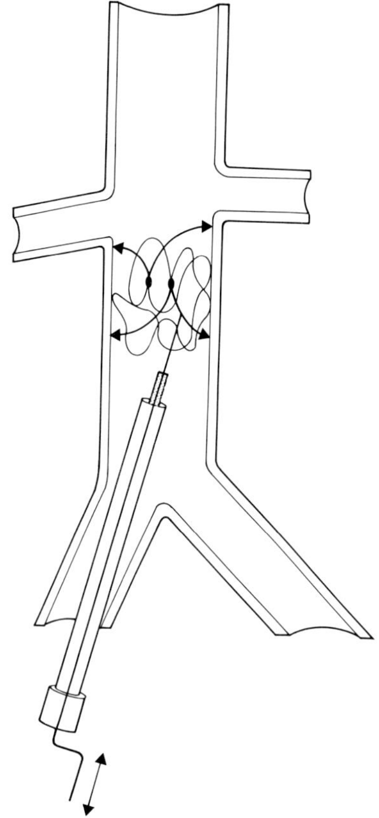

Fig. 5.29
Having withdrawn the catheter to the right angle handle, gentle to and fro motion is used to anchor the lower struts.

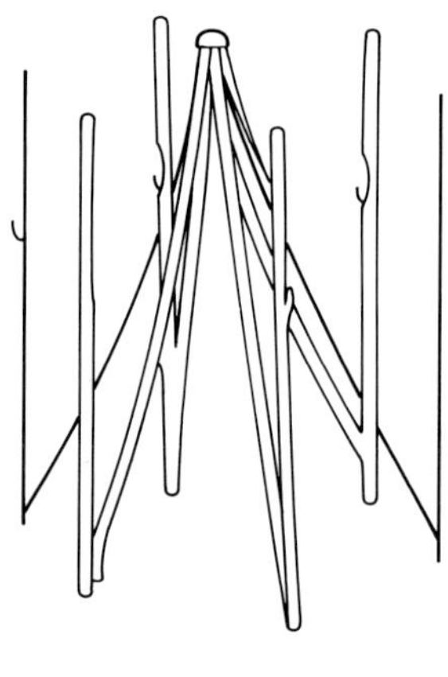

Fig. 5.30
The LGM filter is made up of flat legs with vertical barbed flanges which are designed to help achieve an untilted position in the cava.

11 To disengage the filter the wire pusher handle is turned anticlockwise while pulling gently on it. After 15–20 turns the filter will disengage and the pusher can be removed while continuing to turn it anticlockwise.

12 The filter catheter is now removed and the sheath is removed after a check cavogram is performed if this is felt necessary.

It has been suggested that placement of a BNF from the left femoral vein may be facilitated by using the jugular vein insertion set (Taylor and Roehm, 1991). This is because the femoral sheath may be too short for use through the longer left iliac veins. However, this has not been found to be a problem in most series.

To insert a jugular BNF the steps are essentially similar apart from the positioning procedure which will reflect the fact that the lower struts are placed first followed by the mesh and then the upper struts.

LGM (or Venatech) filter

SPECIFICATION

The LGM filter is manufactured in France by Celsa LG and sold in the USA as the Venatech filter (Ricco *et al.*, 1988). It costs £610. Its design is similar to that of the Greenfield filter but the six struts are made of flat metal strips and there are vertical flanges with small barbs designed to help positioning of the filter in an untilted position within the IVC (Fig. 5.30).

INSERTION TECHNIQUE

Excellent instructions accompany the LGM filter and these should be followed closely. The set comprises a 0.038 inch wire, a 12 Fr dilator/sheath, a blunt implantation cannula and a preloaded filter syringe (Fig. 5.31). A unique feature of the set is that the syringe is supplied with two pistons which are designed to be connected at different ends of the syringe. Depending on whether a femoral or jugular approach is used one or the other of these pistons is selected thereby delivering the filter appropriately orientated and obviating the need for different sets to be purchased (Fig. 5.32).

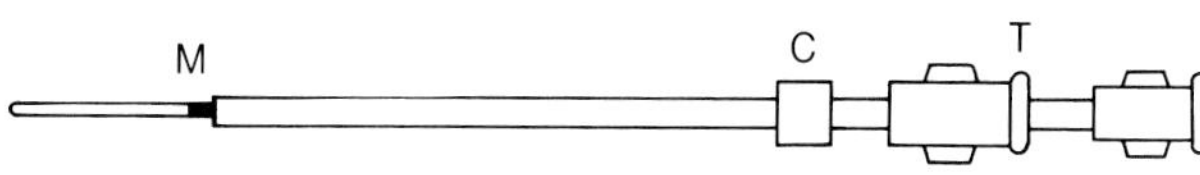

Fig. 5.31
The LGM dilator/sheath has a metal marker (M) on the dilator, a moveable collar on the sheath (C) and a threaded hub on the sheath (T) to take the filter syringe.

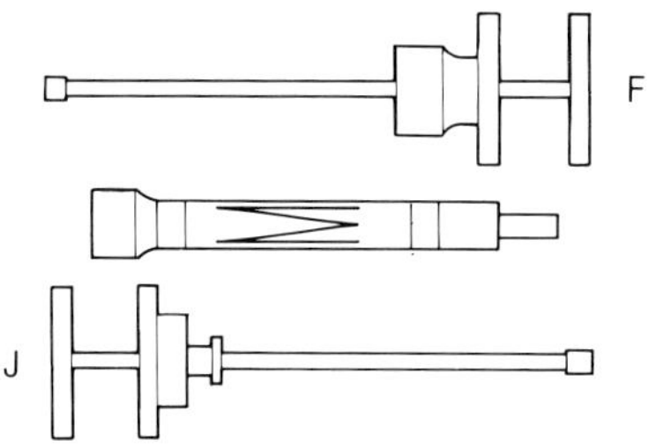

Fig. 5.32
The femoral plunger (F) and the jugular plunger (J) will only fit into the correct end of the filter syringe to ensure correct filter orientation.

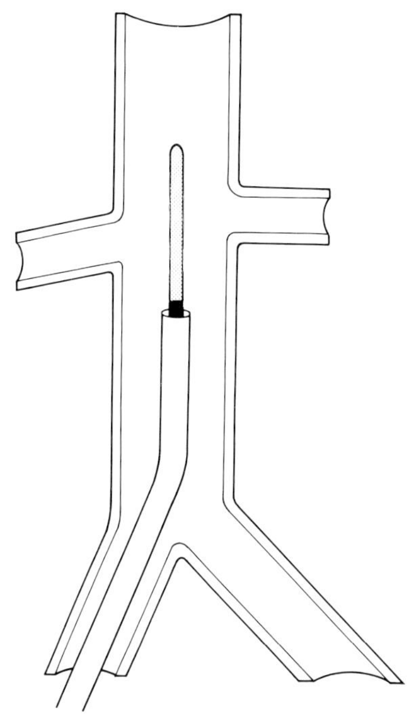

Fig. 5.33
Metallic marker on dilator is positioned 0.5 cm below the renal veins, keeping the dilator fully pushed into the sheath.

To insert a femoral LGM the following steps will be required:

1 Insert a 0.038 inch wire and catheter and perform a cavogram. The LGM should not be used in inferior vena cavas of diameters greater than 28 mm.
2 Determine the level of the desired filter position and relate it to skin markers (Fig. 5.10).
3 Replace the diagnostic catheter with the sheath/dilator and advance them until the metallic marker on the dilator, which represents the end of the sheath, lies 0.5 cm below the renal veins (Fig. 5.33).
4 While holding the sheath/dilator still move the flexible collar along the sheath to the skin to allow subsequent verification that the sheath has not moved.
5 Connect the femoral piston (red) to the filter syringe and having flushed the sheath with heparinized saline screw the syringe to it.
6 Inject the filter into the sheath and then replace the filter syringe with a syringe of heparinized saline and flush the filter a little further into the sheath.
7 Remove the syringe and push the blunt cannula into the sheath until the marker on the cannula reaches the hub of the sheath (Fig. 5.34). The filter will now be at the end of the sheath.
8 The filter is released by holding the blunt cannula absolutely still and quickly pulling the sheath back over it. It is very important to keep the cannula in position as the sheath is withdrawn otherwise the filter will be malpositioned.
9 The cannula and sheath are now withdrawn following a cavogram performed through the sheath if this is felt to be necessary.

To insert a jugular LGM the most important differences are to use the blue jugular piston on the filter syringe and to position the metallic marker on the introducer dilator 4.5 cm below the renal veins.

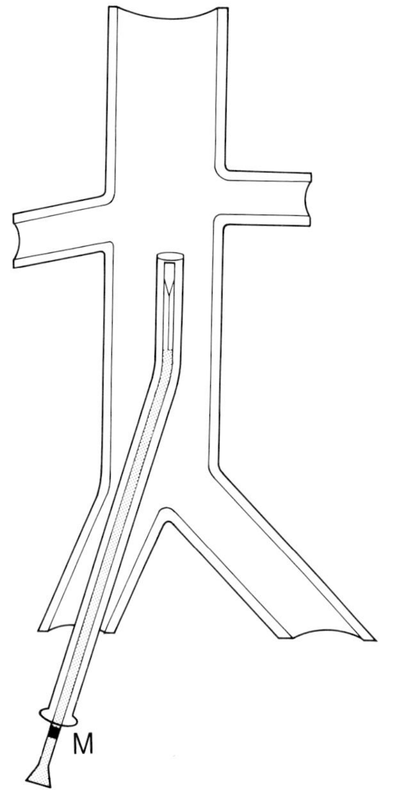

Fig. 5.34
With the marker on the blunt cannula (M) pushed to the hub of the sheath the filter will be at the end of the sheath.

Other filter types

Several IVC filters not already described are under trial.

The Simon nitinol filter is being used in the USA and has the advantage of being introduced through a 9 Fr sheath (Simon *et al.*, 1989). It is constructed of a thermal memory alloy of nickel and titanium which allows the cooled filter to be inserted as straight wires which form the filter shape when warmed to body temperature in the IVC. Early clinical experience suggests good performance although caval occlusion is higher than with other designs (Hawkins and Al-Kutoubi, 1992). The Nitinol Filter is now available from Litechnica Ltd in the UK for a price of £685.

The Amplatz filter has been used quite extensively in the USA but is not available in the UK. It has however been associated with a 23% caval thrombosis rate (McCowan *et al.*, 1990). It requires insertion through a 14 Fr sheath and appears to offer no particular advantages over other currently available filters.

Temporary filtration may be undertaken using the Gunther temporary filter (Cook Inc.). This filter is a modification of the Gunther filter which was used extensively for permanent filtration before being withdrawn following problems with structural integrity. The temporary version is introduced on the end of a guidewire and then withdrawn after 4 to 5 days through a 22 Fr sheath if any emboli have been trapped or through the 6.5 Fr sheath used for introduction if it is empty. There is also a temporary version of the LGM filter.

The length of time that a temporary filter can be left in place will depend on the rate of endothelialization of the filter where it contacts the caval wall and this can be prolonged by moving the position of the filter slightly every day. Careful attention should be paid to sterile technique to minimize the risk of infection around the femoral vein insertion site.

Temporary filter insertion has been recommended in the setting of extensive deep venous thrombosis being treated by thrombolysis. It has also been recommended as prophylaxis against embolism in pelvic and hip surgery where the risk of deep venous thrombosis is high but there is a short-term contraindication to anticoagulation. Until data is available about the incidence of trapped clot in particular clinical settings and the rate of complications associated with the filter the exact role of temporary filtration will remain unclear.

Technical problems: how to overcome them

There are several technical problems which may occur during filter introduction regardless of filter type. These can usually be avoided by paying close attention to the manufacturer's instructions and spending time familiarizing oneself with the workings of the components for introduction at the beginning of the procedure, but advice on how to overcome some of them will now be given.

Filter will not pass through sheath

With the exception of the Bird's Nest filter all the filters described in detail above are fairly rigid along their lengths and any kinking of their delivery sheaths will prevent passage of these filters through the sheath (Fig. 5.35). Kinking of the sheath is most likely to occur if the veins are tortuous which is typically the case with the left iliac vein. For this

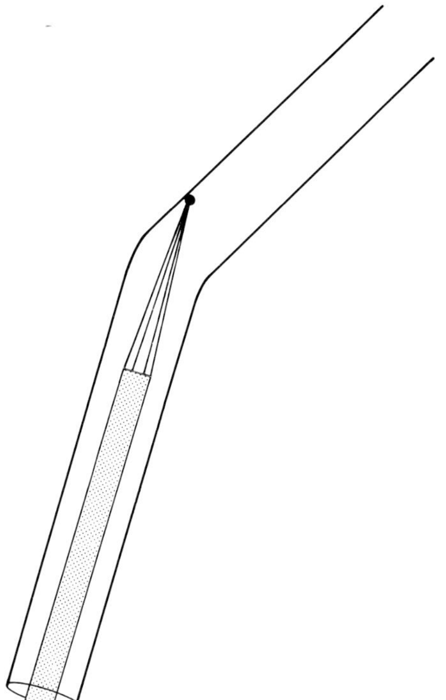

Fig. 5.35
Kinking of the sheath may prevent passage of filter through it.

reason some manufacturers do not recommend the left femoral approach but despite this some radiologists will prefer to try this route before resorting to the jugular vein. If difficulty is encountered when passing the filter through the sheath:

1 Do not push the filter against resistance as this may cause perforation of the sheath.
2 Withdraw the sheath and filter as a unit for a few centimetres and then try advancing the sheath and filter together while rotating the sheath. If this is successful remember to check the position of the tip of the sheath prior to releasing the filter.
3 If there is still resistance either change to the jugular route or try a Bird's Nest filter.

Filter fails to open properly

This will usually occur only if clot has formed in the filter during insertion. This should be prevented by paying close attention to flushing with heparinized saline.

Filter legs asymmetrical in the inferior vena cava

This usually occurs because one or more of the legs has entered a tributary vein. Only large veins seem to cause a problem and careful scrutiny of the cavogram at the beginning of the procedure should prevent this.

Filter misplaced

Filters have been misplaced in the common iliac veins, too high in the IVC and, when approaching from the jugular, in the hepatic, renal and even testicular veins. To prevent these mishaps:

1 Use metal markers on the skin and relate them accurately to the renal veins and iliac confluence.
2 Position the introducer sheath precisely according to the manufacturer's instructions and do not move it during other manoeuvres without repositioning.
3 Use fluoroscopic guidance at all times injecting contrast medium if necessary for localization.

Once a filter has been misplaced it does not usually need to be removed but a second, properly placed filter may be required to establish adequate filtration.

Filter ejected prematurely into the sheath

This will occur with the Titanium Greenfield and Cardial filters if care is not taken to ensure that the release mechanisms are kept locked until release is required. It may also occur with the Cardial filter if resistance is encountered during passage of the filter through the sheath and with the Titanium Greenfield filter if the sheath is not fully locked to the introducer catheter.

COMPLICATIONS OF INFERIOR VENA CAVA FILTRATION

Complications associated with caval filtration are described with all types of filter and in this section general consideration will be given to these complications. Some indication of the complication rates associated with individual filter types will be given in the next section on Results. Table 5.7 lists the main complications of IVC filters.

Femoral vein thrombosis
Bleeding at puncture site
IVC thrombosis
Recurrent pulmonary emboli
Filter migration
IVC perforation
Venous stasis

Table 5.7 *Complications of inferior vena cava (IVC) filtration*

Femoral vein thrombosis

This problem occurred largely as a result of dilating large tracks to allow percutaneous insertion of the stainless steel Greenfield filter when as many as 24% of veins were affected (Mewissen *et al.*, 1989). The smaller sheath sizes now being used rarely lead to this complication.

Bleeding at puncture site

The relatively large size of the introducer sheaths and the possibility of disordered clotting due to anti-coagulation may lead to some prolonged attempts at haemostasis after filtration. In practice however this is usually not a major problem and the incidence of large haematoma formation is very low.

Inferior vena cava thrombosis

This may occur with any filter type and may be due to primary thrombosis or occlusion following clot trapping. It may lead to symptoms of venous stasis and recurrent embolism. The latter occurs either as a result of clot propagation above the filter with subsequent embolization or due to passage of emboli through large collaterals which open up following the thrombosis. The incidence of thrombosis is below 10% with most filters and can be minimized by anticoagulation for several months after filtration if this is not contraindicated. Furthermore, if thrombosis does occur in early follow-up thrombolysis might be considered as a means of removing the thrombus. This will not of course be possible if there is a continuing contraindication to anticoagulation. It should be stressed that most IVC thromboses do not lead to recurrent embolism and symptoms of venous stasis are not invariable (Irving, 1987).

Recurrent pulmonary emboli

This may occur following caval thrombosis as outlined above, but may also be seen if the clot trapping efficiency of the filter is suboptimal (Geisinger *et al.*, 1987). The clot trapping efficiency of filters has been investigated experimentally (Katsamouris *et al.*, 1988) but some doubt has been cast on the relevance of this to the clinical situation. It has been suggested that the Greenfield type of filters lose some of their clot trapping efficiency if they are positioned with a tilt. However, this is not always possible to avoid and the incidence of clinically significant emboli following filtration is very low regardless of filter type.

Filter migration

This may either be distal or proximal. Distal migration is not infrequently seen with some filter designs in the weeks following insertion. It is usually only a few millimetres and unless it is sufficient to bring the filter into an iliac vein it is not of any consequence.

Proximal migration may be much more serious although even migration to the heart does not always lead to symptoms (Castaneda *et al.*, 1983; Friedell *et al.*, 1986). To some extent filter design is a compromise between safe fixation with the risk of damage to the caval wall and atraumatic fixation with the risk of migration. The modern filters have a very low incidence of proximal migration however. It should be remembered that incomplete filter opening, due to clot formation as a result of inadequate flushing during insertion, may lead to migration. It should also be remembered that the radiologist may be able to retrieve a migrated filter using standard retrieval techniques (McCowan *et al.*, 1988; Yakes, 1988). This will involve snaring the filter in such a way as to pull the hooks together to minimize trauma to the heart or veins. It may be possible to reposition the filter in the IVC but if removal is required it will be necessary to dilate the venous access site to allow the insertion of a 24 Fr sheath through which the filter can be extracted. Radiological removal of a filter which has migrated should only be attempted in the first few days following migration, and the risks of attempting removal should be balanced against the risks of leaving the filter where it is.

Inferior vena cava perforation

Perforation of the caval wall by the anchoring struts or legs has been reported with most filter designs. It was a particular problem with early versions of the Titanium Greenfield filter which were prone to splaying with consequent movement of the legs through the wall (Ramchandani *et al.*, 1990), but this problem has now been corrected. Perforation is not usually associated with significant haemorrhage but resultant pericaval fibrosis may narrow the cava and predispose to thrombosis and may produce back pain. Rarely the filter hooks have penetrated other organs.

Venous stasis

Many patients who have IVC filters inserted have extensive deep vein thrombosis which alone can produce symptoms of stasis such as swelling and ulceration. There is no doubt, however, that filter insertion increases the incidence of these complications even when not associated with caval thrombosis. It has been suggested that caval stenosis occurring after pericaval fibrosis may be sufficient to cause stasis (Irving, 1987). Filters which have caused caval occlusion are most likely to be implicated however and there have been reports of fatal phlegmasia cerulea dolens following filter-induced caval thrombosis (Aruny and Kandarpa, 1990).

RESULTS AND CHOICE OF FILTER

An ideal IVC filter would conform to the criteria listed in Table 5.8. No such filter yet exists and probably never will. Each of the filters described in this chapter are reasonably efficient at trapping clot, particularly the larger most clinically significant clots (Katsamouris *et al.*, 1988). With regard to the other criteria in Table 5.8 all the filters are a compromise and each has its supporters. Table 5.9 lists some facts relating to each filter but it should be born in mind that some of the data is based on relatively small series.

High clot trapping efficiency
No migration
No impedance of blood flow
Easy to insert
Stable in the long term
Cheap

Table 5.8 *Characteristics of an ideal inferior vena cava filter*

The rates of caval thrombosis and recurrent embolism given in Table 5.9 refer to reports of clinically significant events. Radiological evidence of occlusion and recurrent embolism may be found in higher numbers but no significant difference in the results of these new filters has emerged in the literature so far (Grassi, 1991). The Bird's Nest filter is the cheapest and has the significant advantage of being suitable for use from both femoral veins. It is also usable in cavas of larger diameter than the other filters. For those departments who wish to keep only one type of filter on their shelves the Bird's Nest filter would be suitable although both femoral and jugular sets would be required. The LGM filter has

Filter type	Cost	Recurrent embolism rate (%)	Caval thrombosis rate (%)	Limit of caval diameter allowed (mm)	Left femoral insertion allowed	Femoral and jugular sets needed
Greenfield Titanium	£787	4.2	0.0	28	No	Yes
Cardial	£590	2.4	7.2	28	No	Yes
Bird's Nest	£420	2.7	2.9	40	Yes	Yes
LGM	£610	2.0	7.0	28	No	No

Table 5.9 *Comparison of some filter parameters*

the advantage of allowing either approach with the same set and like the other filters is slightly simpler to introduce than the Bird's Nest filter. The poor instructions with the Cardial filter will limit its usefulness to the inexperienced operator. The Titanium Greenfield filter has the advantage of being closest in design to the original Greenfield filter about which there is much follow-up data suggesting its long-term efficacy. It is however expensive.

In summary, whichever filter is chosen, the following important points should be born in mind in each case:

1 Assess the balance of risks of filtering versus not filtering and discuss with referring clinician and patient.
2 Read, understand and follow the manufacturer's instructions when inserting filters.
3 Follow-up patients with IVC filters so that our experience broadens and report complications.

Inferior vena cava filtration can be a life-saving technique and the vascular radiologist should keep up to date with developments and make sure his or her colleagues in other disciplines are aware of its potential.

References

ARUNY JE, KANDARPA K (1990) Phlegmasia cerulea dolens, a complication after placement of a Bird's Nest vena cava filter. *American Journal of Roentgenology* **154:** 1105–6.

BARRITT DW, JORDAN SC (1960) Anticoagulant drugs in the treatment of pulmonary embolism. *Lancet* **i:** 1309–12.

BAXTER GM, MACKECHMIE S, DUFFY P (1990) Colour Doppler ultrasound in deep venous thrombosis: A comparison with venography. *Clinical Radiology* **42:** 32–6.

CASTANEDA F, HERRERA M, CARGG AH, SALAMONOWITZ E, LUND G, CASTANEDA-ZUNIGA WR, AMPLATZ K (1983) Migration of a Kimray-Greenfield filter to the right ventricle. *Radiology* **149:** 690.

CHUANG VP, MENA CE, HOSKINS PA (1974) Congenital anomalies of the inferior vena cava. Review of embryogenesis and presentation of a simplified classification. *British Journal of Radiology* **47:** 206–13.

COLEMAN C (1986) Overview of interruption of the inferior vena cava. *Seminars in Interventional Radiology* **3:** 175–88.

DALEN JE, ALPERT JS (1975) Natural history of pulmonary embolism. *Progress of Cardio-Vascular Diseases* **17:** 259–70.

DENNY DF, CRONAN JJ, DORFMAN GS, ESPLIN C (1985) Percutaneous Kimray-Greenfield filter placement by femoral vein puncture. *American Journal of Roentgenology* **145:** 827–9.

FRIEDELL ML, GOLDENKRANZ RJ, PARSONNET V, ALPERT J, BRIEF DK, BRENER BJ, AUERON FM (1986) Migration of a Greenfield filter to the pulmonary artery: A case report. *Journal of Vascular Surgery* **3:** 929–31.

GEISINGER MA, ZELCH MG, RISIUS B (1987) Recurrent pulmonary embolism after Greenfield filter placement. *Radiology* **165:** 383–4.

GRASSI CJ (1991) Inferior vena caval filters: Analysis of five currently available devices. *American Journal of Roentgenology* **156:** 813–21.

GREENFIELD LJ, MCCURDY JR, BROWN PP, ELKINS RC (1973) A new intracaval filter permitting continued flow and resolution of emboli. *Surgery* **73:** 599–606.

HAWKINS SP, AI-KUTOUBI A (1992) The Simon Nitinol inferior vena cava filter: Preliminary experience in the UK. *Clinical Radiology* **46:** 378–350.

IRVING JJ (1987) Inferior vena caval filters. In: *Complications in Diagnostic Imaging*, pp. 440–57. Edited by Ansell G, Wilkins R. Blackwell Scientific Publications, Oxford.

KATSAMOURIS AA, WALTMAN AC, DELICHATSIOS MA, ATHANASOULIS CA (1988) Inferior vena cava filters: *In vitro* comparison of clot trapping and flow dynamics. *Radiology* **166:** 361–6.

MANSOUR M, CHANG AE, SINDELAR WF (1985) Interruption of the inferior vena cava for the prevention of recurrent pulmonary embolism. *American Surgeon* **51:** 375–80.

MCCOWAN TC, FERRIS EJ, REIFSTECK JE, LIU GC, BAKER ML (1988) Retrieval of dislodged Bird's Nest inferior vena caval filters. *Journal of Interventional Radiology* **3:** 179–83.

MCCOWAN TC, FERRIS EJ, CARVER DK, HARSHFIELD DL (1990) Use of external jugular vein as a route for percutaneous inferior vena caval filter placement. *Radiology* **176:** 527–30.

MCCOWAN TC, FERRIS EJ, CARVER DK, BAKER ML (1990) Amplatz vena caval filter: Clinical experience in 30 patients. *American Journal of Roentgenology* **155:** 177–81.

MEWISSEN MW, ERICKSON SJ, FOLEY WD, LIPCHIK EO, OLSON DL, MCCANN KM, SHREIBER ER (1989) Thrombosis at venous insertion sites after inferior vena caval filter placement. *Radiology* **173:** 155–7.

ORSINI RA, JARRELL BE (1984) Suprarenal placement of vena caval filters: Indications, techniques and results. *Journal of Vascular Surgery* **1:** 124–35.

PRINCE MR, NOVELLINE RA, ATHANASOULIS CA, SIMON M (1983) The diameter of the inferior vena cava and its implications for the use of vena caval filters. *Radiology* **149:** 687–9.

RAMCHANDANI P, KOOLPE HA, ZEIT RM (1990) Splaying of Titanium Greenfield inferior vena caval filter. *American Journal of Roentgenology* **155:** 1103–4.

RICCO JB, CROCHET D, SEBILOTTE P, SERRADIMIGNIA A, LEFEBVRE JM, BOUISSOU E, GESLIN P, VIROT P, VAISLIC C, GALLET M, BIRON Y, LEFANT D, DESMARQ JM, DE LA FAYE D (1988) Percutaneous transvenous caval interruption with the LGM filter: Early results of a multicentre trial. *Annals of Vascular Surgery* **3:** 242–7.

ROEHM JOF, GIANTURCO C, BARTH MH, WRIGHT KC (1984) Percutaneous transcatheter filter for the inferior vena cava: A new device for treatment of patients with pulmonary embolism. *Radiology* **150:** 255–7.

ROEHM JOF, JOHNSRUDE IS, BARTH MH, GIANTURCO C

(1988) The Bird's Nest inferior vena cava filter: Progress report. *Radiology* **168**: 745–9.

SANTOS GH, LANSMAN S (1982) Prevention of pulmonary embolism with use of Mobin-Uddin filter. *New York State Journal of Medicine* **82**: 185–8.

SARDI A, MINKEN SL (1987) The placement of intracaval filters in an anomalous (left-sided) vena cava. *Journal of Vascular Surgery* **6**: 84–6.

SCOTT NORRIS C, GREENFIELD LJ, HERRMANN JB (1985) Free-floating iliofemoral thrombus. A risk of pulmonary embolism. *Archives of Surgery* **120**: 806–8.

SELDINGER SI (1953) Catheter replacement of needle in percutaneous arteriography: A new technique. *Acta Radiologica* **39**: 368–76.

SIMON M, ATHANASOULIS CA, KIM D, STEINBERG FL, PORTER DH, BYSE BH, KLESHINSKI S, GELLER S, ORRON DE, WALTMAN AC (1989) Simon Nitinol inferior vena cava filter: Initial clinical experience. *Radiology* **172**: 99–103.

TAYLOR FC, ROEHM JOF (1991) Left femoral vein approach for the percutaneous placement of the Bird's Nest filter. *Cardiovascular and Interventional Radiology* **14**: 342–4.

TEITELBAUM GP, JONES DL, VAN BREDA A, MATSUMOTO AH, FELLMETH BD, CHESPAK LW, BARTH KH (1989) Vena caval filter splaying: Potential complication of use of the Titanium Greenfield filter. *Radiology* **173**: 809–14.

WELLS IP (1989) Inferior vena cava filters and when to use them. *Clinical Radiology* **40**: 11–12.

YAKES WF (1988) Percutaneous retrieval of Kimray-Greenfield filter from right atrium and placement in inferior vena cava filter. *Radiology* **169**: 849–51.

Index

Aneurysms, false 74
Angiomyolipomas 75
Abscess in PTA 22
Acylated plasminogen streptokinase activator complex 56, 57, 58
Allison fragment grasper 87
Amplaz filter 112
Amplaz superstiff wire 13
Amputation 58
  rates 33
  for thrombosis 37
Angiography
  before embolization 67
Angioplasty
  below the inguinal ligament 19
  complications of site 22
  new devices in 25
  occlusion at site of 22
  popliteal artery thrombosis after 33
  residual thrombus after 54
  rotational transluminal catheter system 25
Anticoagulation
  contraindications to 94
  PTA and 3
  in thrombolysis 55
Aortic bifurcation
  advancing guidewire across 46
  passage of balloon catheter over 14
  saddle embolus of 37
Aortic occlusions
  contraindicating PTA 2
  thrombolysis and 47
Aortoiliac segment
  balloon dilation of 19
Arterial access
  indications for single wall puncture 10
Arterial puncture
  complications of 21
  needles for 4
  peroperative 15
  technique 11
Arterial rupture
  during balloon dilation 22
Arteriography
  in arteriovenous malformations 69
  in embolization 67
  pre-thrombolysis 41
Arteriovenous fistula 74, 75
  complicating PTA 22
Arteriovenous malformations
  balloon embolization 66
  embolization for 62, 68, 69, 70, 71, 72
  gluteal 71, 72
  of hand 72
  pelvic 73
  presentation of 68
  pulmonary 66
  surgery for 68, 69
Aspiration thrombectomy 2, 23
Atheroma
  laster ablation 26
Axillary artery puncture 14, 21

Balloon
  diameters 18
  for embolization 66
Balloon catheters 4
  features of 4
  inflation devices 6
  types of 6
Balloon dilation 18
  adequate 19
  arterial rupture during 22
Bentsen guide wires 4, 5, 16
Bird Nest filter 100, 107, 112, 115
Bleeding disorders
  contraindicating vena cava filtriation 95
Blood vessels
  perforation of 91
  trauma to *See Vascular injury*
Blue toe syndrome 2
Bone tumours
  embolization 75, 77
Brachial artery puncture 14
Brachial plexus
  care of 14

Bucrylate 66
By-pass grafts
  puncturing 15

Cardiac arrhythmias 91
Cardial filter 100, 105, 113, 115, 116
Catheter(s)
  accidental removal of 48
  balloon *See Balloon catheters*
  curved 17
  for delivering embolization coils 65
  femorovisceaeral 13
  guiding and predilating 4
  headhunter (femorocerebral) 63
  introducer sheaths 6, 7
  introducer for foreign body retrieval 84
  pigtail/preshaped 86, 87, 99
  preshaped curved 5, 12
  sidewinder 13, 63
  Swan-Ganz 86
  tip shapes 5
  trackability of 6
Cerebrovascular accidents
  in thrombolysis 53
Chemoembolization 66
Cholesterol embolization 24
Cimino fistulae
  dilation in 20
Clotting disorders
  contraindicating thrombolysis 40
Cobra catheter 63
Common femoral artery
  antegrade catheterization 44
Common femoral artery embolism 32, 33
  extending into superficial artery 38
Common femoral artery occlusion 41
Common peroneal arteries
  occlusions in 46
Contralateral common femoral artery
  as site for PTA 8, 13
  puncture 13
Coronary stent
  misplaced 89

Deep venous thrombosis
  with contraindication to anticoagulation 94
  with pulmonary hypertension 95
  thrombolysis in 112
Dementia 40
Diabetes
  PTA in 3
Digital subtraction angiography 18, 63
Dissection
  during recanalization 17
  in PTA 22
Dormia baskets 87, 88

Embolectomy
  after PTA 24
Embolic agents 64
  choice of 68
  liquid 66
  mechanical 65
  reaction to 79
Embolization after PTA 23
Embolization coils 65, 69, 73
  retrieval when misplaced 90

Embolization, therapeutic 61–80
  agents for 64, 65, 66, 68
  anaesthesia for 67
  angiography in 67
  for arteriovenous malformations 68
  avoidance for complications 78
  for bone tumours 75, 77
  catheters for 63
  choice of agent 68
  complications 77
  death in 79
  delayed complications 79
  equipment for 62, 63
  gas formation in 79
  guidewires for 63
  indications 62
  infection following 77, 79
  informed consent for 67
  materials for 62
  paraesthesia following 79
  personnel for 62
  premedication in 67
  renal failure following 79
  retrograde thrombosis following 79
  surgical cover for 67
  technique for 67
  tissue necrosis in 79
  for tumours 75
  for vascular injury 74
Epistaxis 52
Ethanol in embolization 66
External iliac artery thrombosis 42

Femoral arteries
  See also Superficial femoral arteries etc
  passing wire 10
Femoral artery insertion
  of vena cava filter 102, 105, 107, 111
Femoral artery occlusions
  thrombolysis for 43
Femoral artery puncture
  angle of needle 9
  complications of 21
  difficult 44
Femoral artery stenosis 54
Femoral artery thrombosis 31, 32, 49
Femoral vein cannulation
  in vena cava filtration 97, 98
Femoral vein thrombosis
  in vena cava filtration 114
Femorodistal vein graft thrombosis 35
Femoro-femoral graft thrombosis 32
Femoro-popliteal graft thrombosis 31
Femoropopliteal occlusion 18
Femoropopliteal percutaneous transluminal angioplasty 25
Femorovisceral catheter 13
Fibrinogen-fibrin system 56
Fibrinolytic drugs 47, 55
Fibrinolytic system 56
Fibrinolytic therapy 30
  for ischaemic episodes 2
Forceps
  for foreign body retrieval 88
Foreign body retrieval 81–92
  adjunctive methods 88

arrhythmias during 91
balloon methods 88
catheter introducer sheath set 84
complications of 89
contraindications 83
equipment for 84
failure 89
forceps for 88
fragmentation during 89
indications for 82
infection following 91
loop snares for 86
patient preparation 84
postprocedural care 91
premedication 84
sites of 82
snaring 87, 88, 89
technique of 84, 85
types of 82
vascular perforation in 91

Gastrointestinal haemorrhage 52
Gelatin sponge as embolic agent 65, 69
Gianturco-Roehm bird's nest filter 100, 107, 112, 115
Gluteal arteriovenous malformation 71, 72
Graft fistulae 20
Graft occlusion
thrombolysis for 47
Graft thrombectomy 33
Greenfilters 113, 114, 116
Groin haematomas
in thrombolysis 48
Gruntzig balloon catheters 4, 6
Guidewires 4
dissection 17, 18
extrastiff 19
for foreign body retrieval 86
J curves 63
Newton 63
passage of 7
passing aortic bifurcation 46
Rosen 63
Terumo 64
types of 5
Gunther temporary filter 112

Haemangiomas 75
Haematemesis 52
Haematoma
abscess formation in 22
after embolization for tumours 77
avoidance of 14, 15, 21, 54
incidence of 22
management of 21
risk of 44
in thrombolysis 48, 52, 57
Haematuria 52
Haemoptysis 52
Hand
arteriovenous malformations of 72
Headhunter catheter 63
Heparin
PTA and 3
in thrombolysis 55
Hypertension
contraindicating thrombolysis 40

Hypertonic dextrose
in embolization 66, 69

Iliac artery
foreign body in 90
Iliac artery occlusions 41
contraindicating PTA 2
Iliac artery stenosis 54
Iliac artery thrombosis 31, 32, 42
Infection
following embolization 77, 78, 79
following retrieval of foreign body 91
Inferior vena cava
congenital variations in 99
diameter of 99
Inferior vena cavogram 99
Inferior vena cava filtration 93–117
Amplaz filter 112
bleeding at puncture site 114
cardial filter 100, 105, 113, 116
characteristics of filters 114
choice of filter 96, 115
clotting parameters and 96
complications of 114
consent to 96
contraindications to 95
failure of filter 113
filter migration 114
filter misplaced 113
filter types 112
Gianturco-Roehm Bird's Nest filter 100, 107, 112, 115
Greenfield filters 113, 114, 116
Gunther temporary filter 112
indication for 94
insertion technique, 97, 102–12
Kimray-Greenfield filter 100, 101, 106, 113
LGM filter 110
life of filters 95
materials for 100
Mobin-Uddin umbrella 100
perforation in 115
postinsertion routine 100
precautions 96
premature ejection of filter 113
preparation for 96
results of 115
Simon nitinol filter 112
site of insertion 96
skin markers for 102
suprarenal 95
technical problems 112
temporary 112
thrombosis in 114, 115
Titanium Greenfield filter 100, 101, 106, 113, 115, 116
Venatech filter 110
venous stasis after 115
Inferior vena cava thrombosis 114
Intermittent claudication 41
Intra-arterial thrombolytic therapy
See under Thrombolysis
Intravascular foreign bodies 82
See also Foreign body retrieval
Ipsilateral femoral artery
access to 8
puncture of 10
for PTA 8

Ischaemia of leg
  See under Leg ischaemia
Isobutyl-2-cyanoacrylate 66

Jakob-Creutzfeld disease 65
Jugular vein
  foreign body in 83
Jugular vein puncture
  for vena cava filtration 97, 98, 103, 104, 107, 110, 111

Kallikrein 56
Kensey catheter 25
Kimray-Greenfield filter 100, 101
Knee
  metastases in 76, 77

Lasers
  in ablation of atheroma 26
Leg fractures 40
Leg ischaemia
  conservative management 40
  deterioration 53
  indications for surgery 37
  thrombolytic therapy for 30
LGM filter 110, 115
Liver
  metastases in 77
Loop snares 89
  for foreign body retrieval 85
Lower limb ischaemia *See Leg ischaemia*
Lung
  arteriovenous malformations in 66
Lyophylized human dura as embolic agent 65

Marsmann wire 66
Median nerve
  care of 14
Melaena 52
Metastases
  embolisation for 76, 77
Mobin-Uddin umbrella 100

Newton cerebral guide wires 4, 5, 16, 63

Occlusion
  from stenosis 22
  recanalization 17
  stenosis proximal to 46
  steps in crossing 17
Olbert ballon catheters 4, 6

Pain following embolization 78
Parasthesia following embolization 79
Pelvic arteriovenous malformation 73
Percutaneous transluminal angioplasty 1–27
  abscess complicating 22
  anticoagulant therapy and 3
  arterial access 8
    See also under specific arteries
  arteriovenous fistula complicating 22
  below inguinal ligament 19
  complications of 21
  consent for 3
  contraindications for 2
  in diabetes 3
  distal site complications 23
  embolization following 23
  equipment for 3
    See also under items of equipment
  femoropopliteal 25
  haemorrhage during 22
  indications 2
  indications for arterial access 8
  neural damage in 22
  patient preparation 2
  premedication for 3
  pseudoaneurysm and thrombosis in 22
  postprocedural care 21
  puncture site complications 21
  recanalization technique 16–20
  of renal dialysis fistula 20
  results of 24
  sites of access 8
  subclavian 20
  systemic complications 24
  X-ray equipment for 6
Peripheral vascular disease
  grading 2
Peroperative arterial puncture 15
Phlegmasia cerula dolens 115
Pigtail/preshaped catheters 86, 87, 99
Plasmin 55
Plasminogen activators 56
Plasminogen-plasmin system 56
Platelet activating factor 56
Polyvinyl alcohol as embolic agent 65
Popliteal artery
  foreign body in 89
  puncture of 15
  thrombolysis 50, 51, 52
Popliteal artery embolus 31
  extending into tibial artery 45, 51, 52
Popliteal artery occlusion
  streptokinase infusions in 30
  thrombolysis for 43
Popliteal artery stenosis
  difficulty in thrombolysis 54
Popliteal artery thrombosis 43
  after angioplasty 33
  extending into tibial arteries 34, 50
Popliteal fossa
  relationships in 15
Postembolization syndrome 77, 78
Post pulmonary embolectomy 95
Pregnancy
  contraindicating thrombolysis 40
Pressure measurements
  in recanalization 19
Profunda femoris artery
  puncture of 11, 12, 15
Pseudoaneurysm complicating PTA 22
Pulmonary embolism
  anticoagulation contraindicated 94
  complicating therapeutic embolization 78
  death from 94
  recurrent 94, 114, 115
  risk of 94
Pulmonary hypertension
  deep venous thrombosis with 95
Pulsed spray pharmaco-mechanical thrombolysis 59

Radiography
for therapeutic embolization 62
Recanalization technique 16–20
balloon dilation 18, 19
complete occlusions 17
dissection during 17
pressure measurements in 19
stenosis 16
stenting after 26
Recombinent tissue plasminogen activator 56
Rectal bleeding 52
Renal ablation 66
Renal dialysis fistula
PTA of 20
Renal failure 58, 79
Rest pain 41
Retroperitoneal haemorrhage in thrombolysis 53
Rosen guidewires 13, 63
Rotational transluminal angioplasty catheter system 25

Sidewinder catheter 13, 63
Simon nitinol filter 112
Single chain urokinase plasminogen activator 56, 57, 59
Smart needle 15
Steal phenomenon 69
Steerable wires 5
Stenosis
converting into occlusion 22
proximal to occlusion 46
recanalization technique 16
steps in crossing 16
Stiff wire guides 5
Streptokinase 30, 47, 56
allertic reations to 58
dosage 59
results of treatment 58
Subclavian percutaneous transluminal angioplasty 20
Superficial femoral artery puncture 11, 12, 15
Superficial femoral artery embolus 34
Superficial femoral artery occlusion 41
extending into popliteal artery 43
Superficial femoral artery thrombosis 31, 32, 39
Superficial femoral/popliteal artery occlusion 36
Superior vena cava
foreign body retrieval from 84, 85, 86
Superior vena cava syndrome 26
Superstiff wires 13
Suprarenal inferior vena cava filtration 95
Surgical thromboembolectomy 37
Swan-Ganz catheter 86
Swelling following embolization 79

Terumo glide wire 4, 5, 16, 64
Therapeutic embolization
See Embolization
Thromboembolectomy
surgical 37
Thrombolysis 24, 29–60
arterial perforation in 52
blood tests in 55
catheter kinking 48
catheter removal 48
cerebrovascular accident in 53
complications of 57
contraindications to 40
criteria for 58
in deep venous thrombosis 112
distal embolization of thrombus in 49
drugs used 47
fibrinolytic drugs in 55
gastrointestinal haemorrhage in 52
haematological management 55
indications for 30
informed consent for 41
infusion in regime 47, 59
in long occlusions 46
management of 47
manipulation of catheter 46
observations during 47
oozing from puncture site 46
pericatheter thrombosis 49
postlysis management 53, 54
prelysis management 40
prelysis problems 44
problems in 44, 48, 54
pulsed spray pharmaco-mechnical 59
radiology in 40
recent advances 58
residual stenosis after 53
results of 58
retroperitoneal haemorrhage in 53
risks of 40
slow blood flow after 55
systemic complications 57
thrombosis after 57
Thrombosis
retrograde 79
thrombolysis for
See under Thrombolysis
Tibial artery occlusion
thrombolysis for 43, 46
Tibial artery thrombosis 45
Tibial fracture 40
Tibial percutaneous transluminal angioplasty 25
Tissue necrosis in embolization 79
Tissue plasminogen activator 47, 57, 59
Titanium Greenfield filter 100, 101, 106, 115
Transluminal extraction catheter 25
Trauma to blood vessels
See Vascular injury
Tumours
definitive embolization 75
embolization for 75
Tuohy Borst adaptor 7

Urokinase 47, 56
dosage 59
results of treatment 58

Van Andel predilating catheter 4, 19
Vascular foreign bodies
embolization of 91
fragmentation of 89
risks of 83
Vascular injury
acute 74
delayed complications 74
embolization for 74
Vascular surgery 37
Venatech filter 110

Vena cava
  See under Inferior vena cava and Superior vena cava
Venous stasis
  after vena cava filtration 115
Venous thrombosis
  See under Deep venous thrombosis

Vertrebral artery
  balloon dilation and 20

Warfarin
  PTA and 3
  in thrombolysis 55